STUDY GUIDE
for

Moore and McCabe's
Introduction to the
Practice of Statistics
Fourth Edition

MICHAEL A. FLIGNER
WILLIAM I. NOTZ
Both of Ohio State University

W. H. Freeman and Company
New York

Printed in the United States of America.

ISBN 0-7167-4912-2

Second printing, 2003

CONTENTS

Preface

PREFACE TO THE STUDY GUIDE

We have written this study guide to help you learn and review the material in the text. The structure of this study guide is as follows. We first provide an overview of each section, which reviews the key concepts of that section. After the overview, there are *guided* solutions to selected odd-numbered problems in that section, along with the key concepts from the section required for solving each problem. The guided solutions provide hints for setting up and thinking about the exercise, which should help improve your problem-solving skills. Once you have worked through the guided solution, you can look up the complete solution provided later in the study guide to check your work.

What is the best way to use this study guide? Part of learning involves doing homework problems. In doing a problem, it is best for you to first try to solve the problem on your own. If you are having difficulties, then try to solve the problem using the hints and ideas in the guided solution. If you are still having problems, then you can read through the complete solution. Generally, try to use the complete solution as a way to check your work. Be careful not to confuse reading the complete solutions with doing the problems themselves. This is the same mistake as reading a book about swimming and then believing you are prepared to jump into the deep end of the pool. If you simply read our complete solutions and convince yourself that you could have worked the problem on your own, you may be misleading yourself, which can lead to trouble on exams. When you are having difficulty with a particular type of problem, find similar problems to work in the exercises in the text. Often problems adjacent to each other in the text use the same ideas.

In the overviews we try to summarize the material that is most important, which should help you review material when you are preparing for a test. If any of the terms in the overview are unfamiliar, you probably need to go back to the text and reread the appropriate section. If you are having difficulties with the material, don't neglect to see your instructor for help. Face-to-face communication with your instructor is the best way to clear up difficulties.

We would like to thank Alysha Fligner for her inspiration as we prepared this revision.

CHAPTER 1

LOOKING AT DATA - DISTRIBUTIONS

SECTION 1.1

OVERVIEW

Section 1.1 introduces several methods for exploring data. These methods should only be applied after clearly understanding the background of the data collected. The choice of method depends to some extent upon the type of variable being measured. The two types of variables described in this section are

- **Categorical variables** - variables that record to what group or category an individual belongs. Hair color and gender are examples of categorical variables. Although we might count the number of people in the group with brown hair, we wouldn't think about computing an average hair color for the group, even if numbers were used to represent the hair color categories.

- **Quantitative variables** - variables that have numerical values and with which it makes sense to do arithmetic. Height, weight, and GPA are examples of quantitative variables. It makes sense to talk about the average height or GPA of a group of people.

To summarize the **distribution** of a variable, for categorical variables use **bar charts** or **pie charts**, while for numerical data use **histograms** or **stemplots**. Also, when numerical data are collected over time, in addition to a

1

histogram or stemplot, a **timeplot** can be used to look for interesting features of the data. When examining the data through graphs we should be on the alert for

- unusual values that do not follow the pattern of the rest of the data

- some sense of a central or typical value of the data

- some sense of how spread out or variable the data are

- some sense of the shape of the overall pattern

In addition, when drawing a timeplot be on the lookout for **trends** occurring over time. Although many of the graphs and plots may be drawn by computer, it is still up to you to recognize and interpret the important features of the plots and the information they contain.

GUIDED SOLUTIONS

Exercise 1.3

KEY CONCEPTS - individuals and type of variables

Identify the "individuals" or objects described, then the "variables" or characteristics being measured. Once the variables are identified, you need to determine if they are categorical (the variable just puts individuals into one of several groups) or quantitative (the variable takes meaningful numerical values for which arithmetic operations make sense).

The "individuals" in this problem are the funds. If we included share price, this would be a quantitative variable. If we had another variable, say a 1 if the year to date return was positive and 0 if it wasn't, this would still be a categorical variable even though we used numbers to represent the two categories. Now list the variables recorded and classify each as categorical or quantitative.

Name of Variable   Type of Variable

Exercise 1.7

KEY CONCEPTS - interpreting variables

To determine what a variable is telling us, we must know the purpose for which the variable is to be used. If the purpose is to make comparisons, as here, we need to consider whether the groups being compared differ only in the value of the variable or if they differ in other ways. If the groups differ in other ways, ask yourself if these differences could be partly responsible for differences in the value of the variable. For example, if two groups differ in size, then variables related to size (such as counts), are likely to differ even though the groups are identical in all other respects. In this exercise, consider ways in which the groups (different years) might differ and if these might explain the differences in cancer death rates even if cancer treatments are becoming more effective. Here are some scenarios that can be applied to different parts of the problem.

Suppose that cancer was detected earlier. What would that do to survival rates?

Suppose that population size increases and 1% of the population dies of cancer each year.

Suppose that progress is made on the fight against heart disease. What would be the effect on the death rates due to cancer?

Exercise 1.19

KEY CONCEPTS - interpreting a histogram

How would you describe this distribution? Which portion of the histogram do you think corresponds to the state schools? How about the more exclusive private schools? In general, how many groups of schools are there and what are the most important aspects of the distribution?

Exercise 1.26

KEY CONCEPTS - drawing histograms and stem-and-leaf plots, and interpreting their shapes

DRP scores

```
40  26  39  14  42  18  25  43  46  27  19
47  19  26  35  34  15  44  40  38  31  46
52  25  35  35  33  29  34  41  49  28  52
47  35  48  22  33  41  51  27  14  54  45
```
How to draw a histogram.

a). When drawing a histogram, choose class intervals that divide the data into classes of equal length. For this data set, the smallest DRP score is 14 and the largest is 54, so the class intervals need to cover this entire range. A simple set of class intervals would be 10 - < 20 (10 is included in the interval but not 20), 20 - < 30, 30 - < 40, and 40 - < 50 and 50 - < 60. Other sets of intervals are possible, although these have the advantage of using fairly simple numbers as endpoints.

b). Count the number of data values in each class interval. Using the class intervals above, complete the frequency table below. Remember, when counting the number in each interval be sure to include data values equal to the lower endpoint but not the upper endpoint.

Interval   Count   Percent

c). Draw the histogram, which is a picture of the frequency table. In addition to drawing the bars, this also requires labeling the axes. Using the frequency table you have computed, complete the histogram given. The x axis has been labeled for you. Make sure to include an appropriate label for the y axis, as well as numbers for the scale.

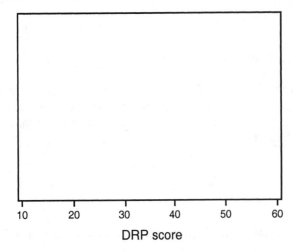

10   20   30   40   50   60

DRP score

How to draw the stem-and-leaf plot - single stems and splitting stems

a). It is easiest although not necessary to first order the data. If the data has been ordered, the leaves on the stems will be in increasing order. The DRP scores have been ordered for you below.

14   14   15   18   19   19   22   25   25   26   26   27   27   28   29   31   33   33   34
34   35   35   35   35   38   39   40   40   41   41   42   43   44   45   46   46   47   47
48   49   51   52   52   54

b). Using the stems below, complete the stem-and-leaf plot.

1|
2|
3|
4|
5|

What are the similarities between this stem-and-leaf plot and the histogram you drew? Think about the relationship between the stems and the class intervals in this example, which helps to explain why the stem-and-leaf plot looks just like the histogram laid on its side. What class intervals do the stems correspond to? What is one important difference between the histogram and the stem-and-leaf plot?

c). Using the DRP scores, increase the number of stems by splitting each of the previous stems in two. Complete the split stem-and-leaf plot in the space given.

In this case, splitting the stems results in a plot similar to a histogram, which uses too many class intervals. (How would you construct a histogram that corresponds to this stem-and-leaf plot? What would the class intervals be?) The more regular features of the data set are becoming obscured by the extra details you are forced to look at with the extra stems. As a display of the data, would you prefer the histogram or the first stem-and-leaf plot? Why?

To finish up the example, think about the important features that describe a distribution. Does the distribution of the DRP scores have a single peak? Does it appear to be symmetric, or is it skewed to the right (tail with larger values is longer), or to the left? Are there any outliers that fall outside the overall pattern of the data?

Exercise 1.37

KEY CONCEPTS - drawing and interpreting a timeplot

a) Complete the timeplot on the graph on the next page. The first three points are plotted for you.

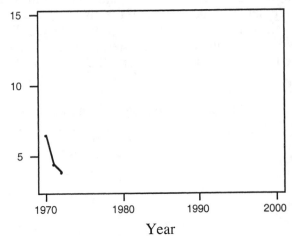

Year

Over the 30 year period plotted, the interest rate shows several clear cycles. These cycles produce three well defined clear temporary peaks which involve more than a single year of increase in the rates. Identify these. There was an overall peak in interest rates in the early 1980s. Has there been a consistent downward trend in rates since that time?

COMPLETE SOLUTIONS

Exercise 1.3

Name of Variable	Type of Variable
Category	Categorical
Net Assets	Quantitative
Year to date return	Quantitative
Largest holding	Categorical

Exercise 1.7

a) As the population size increases and people also live longer, the number of people dying each year from cancer, as well as other causes will go up even if treatments are more effective, simply because there are more people. For example, suppose the number of people at risk for cancer (because of age) increases from 1000 to 1500. Suppose treatments have reduced the incidence of cancer from 5% to 4%. The number of cancer deaths will increase - from 50 to 60!

b) People die of something - if other death rates go down, the cancer death rate could still go up even if cancer treatments were more effective. People are more likely to survive long enough to contract and die from cancer than another disease.

c) Suppose cancer was detected earlier, but treatments were not more effective. People would appear to live longer just due to earlier diagnoses. For example, if breast cancer was being diagnosed one year earlier but life expectancy was not changed by treatment, patients would appear to be surviving one year longer (a woman with breast cancer dies at 55 - she was diagnosed at 52 instead of 53, so her "survival" time is 3 years instead of 2).

Exercise 1.19

There are three groups of schools. The state schools such as the University of Massachusetts have the lower tuitions, and form the group with tuitions of $6000 and below. The remaining private schools seem to be divided into two groups. There are 22 schools in the range $12,000 - $18,000 and include less expensive private schools such as Northeastern University. At the high end of the distribution (over $24,000) are some of the most expensive private colleges which include, for example, Harvard University. This distribution provides an example of a trimodal (three modes) distribution that has been created by including three distinct groups of schools in the distribution.

Exercise 1.23

Interval	Count	Percent
10 - < 20	6	13.63%
20 - < 0	9	20.45%
30 - < 40	11	25.00%
40 - < 50	14	31.82%
50 - < 60	4	9.09%
	44	99.99%

```
Stem-and-leaf plot of DRPscore, with single stems

1 | 445899
2 | 255667789
3 | 13344555589
4 | 00112345667789
5 | 1224
```

```
Stem-and-leaf plot of DRPscore, with split stems
1 | 44
1 | 5899
2 | 2
2 | 55667789
3 | 13344
3 | 555589
4 | 0011234
4 | 5667789
5 | 1224
```

In general, for this number of observations, the preference for a histogram over a stem-and-leaf plot is a personal preference. For smaller data sets, the stem-and-leaf plot is usually preferred, while for larger data sets most people prefer the histogram. In this particular example, the data set is probably starting to get a little large for the stem-and-leaf plot which, by keeping a record of every observation, is beginning to look a little cluttered. So our preference would be a histogram - it seems to make the general shape of the distribution a little more apparent.

Comments on the general shape: There are no obvious outliers that depart from the general pattern. The distribution is unimodal and skewed to the left. Test scores, which have an upper bound on the maximum score that students come close to, are often left-skewed. Although there is a lower bound of zero, the scores generally don't go quite that low, but instead slowly trail off on the lower end, giving the left-skewed appearance.

Exercise 1.37

a)

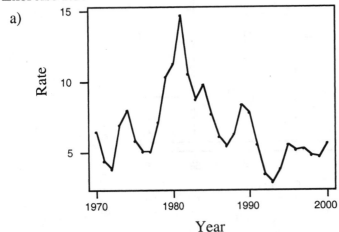

b) The three well defined temporary peaks occurred in 1974, 1981, and 1989. While there is a "peak" in 1984, it is not part of a clear up-and-down movement, nor is the higher interest rate in 1995.

c) The overall peak in the interest rate over these years occurred in 1981. There was a general downward trend from 1981 until around 1993, but since 1993 the rates have come back up slightly and then leveled off.

SECTION 1.2

OVERVIEW

Although graphs give an overall sense of the data, numerical summaries of features of the data make more precise the notions of center and spread.

Two important measures of center are the **mean** and the **median.** If there are n observations, $x_1, x_2, ...,x_n$, then the mean is

$$\bar{x} = \frac{x_1 + x_2 + ... + x_n}{n} = \frac{1}{n}\sum x_i$$

where $\sum$ means "add up all these numbers." Thus, the mean is just the total of all the observations divided by the number of observations.

While the median can be expressed by a formula, it is simpler to describe the rules for finding it.

How to find the median.

1. List all the observations from smallest to largest.

2. If the number of observations is odd, then the median is the middle observation. Count from the bottom of the list of ordered values up to the $(n + 1)/2$ largest observation. This observation is the median.

3. If the number of observations is even, then the median is the average of the two center observations.

The most important measures of spread are the **quartiles,** the **standard deviation**, and **variance.** For measures of spread, the quartiles are appropriate when the median is used as a measure of center. In general, the median and quartiles are more appropriate when outliers are present or when the data are skewed. In addition, the **five-number summary**, which reports the largest and smallest values of the data, the quartiles and the median, provides a compact description of the data that can be represented graphically by a **boxplot**. Computationally, the first quartile, Q_1, is the median of the lower half of the list of ordered observations and the third quartile, Q_3, is the median of the upper half of the list of ordered values.

If you use the mean as a measure of center, then the standard deviation and variance are the appropriate measures of spread. Remember that means and variances can be strongly affected by outliers and are harder to interpret for skewed data.

If we have n observations, x_1, x_2, ...,x_n, with mean $\bar{x}$, then the variance s^2 can be found using the formula

$$s^2 = \frac{(x_1 - \bar{x})^2 + (x_2 - \bar{x})^2 + ... + (x_n - \bar{x})^2}{n-1} = \frac{1}{n-1}\sum(x_i - \bar{x})^2$$

The standard deviation is the square root of the variance, i.e., $s = \sqrt{s^2}$, and is a measure of spread in the same units as the original data. If the observations are in feet, then the standard deviation is in feet as well.

GUIDED SOLUTIONS

Exercise 1.49

KEY CONCEPTS - measures of center, five-number summary, drawing boxplots

a) Complete the back-to-back stemplot using the stems below. There are only a few observations over a fairly wide range, so the overall shapes of the distributions tend to be indistinct.

```
Women        |    | Men
             |  7 |
             |  8 |
             |  9 |
             | 10 |
             | 11 |
             | 12 |
             | 13 |
             | 14 |
             | 15 |
             | 16 |
             | 17 |
             | 18 |
```

b) To find the means, find the sum of the scores, then divide by the number of scores. For the median, since the number of scores is even for both groups, the median is the average of the two middle scores. You can find this easily from the stemplot.

The feature that would suggest that $\bar{x} > M$ is not necessarily the same for both distributions, so look at them carefully.

c) The five-number summary consists of the median, the minimum and the maximum, and the first and third quartiles. Remember that the first quartile is the median of those observations below the median (for the women there are 9 observations below the median) and the third quartile is the median of those observations above the median. Fill in the table below. These numbers are used to draw the boxplot.

	Men	Women
Minimum		
Q_1		
M		
Q_3		
Maximum		

To determine if the $1.5 \times$ IQR criterion flags the largest women's observation, first compute the IQR. It is $Q_3 - Q_1$.

IQR = $1.5 \times$ IQR =

Then add $1.5 \times$ IQR to the third quartile and see if the largest women's observation exceeds this value.

$$Q_3 + (1.5 \times \text{IQR}) =$$

Is 200 larger than this number? If so, then it is flagged as an outlier. We have drawn the boxplot for the men. After making sure you understand the men's boxplot, add the women's boxplot to this picture.

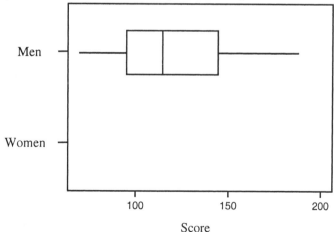

The first thing is to draw the box, which goes from the first quartile to the third quartile. Next locate the median within the box. Finally, check for outliers. If there are no outliers, the lines from the box extend to the smallest and largest observations. If there are outliers, then the lines from the box extend to the smallest and largest observations which are not outliers. The outliers are then identified individually with a symbol (usually either a * or a dot).

Drawing one or two boxplots or other graphical display by hand is the best way to make sure you understand how to interpret the display. But after that, it is really best to leave the drawing of boxplots and most other graphical displays to a statistical computer package.

d) Use both the stem-and-leaf plot and boxplot to answer the questions. Which graphic makes it easier to answer the questions?

Exercise 1.59

KEY CONCEPTS - measures of center

When there are several observations at a single value, the key is to remember that the mean is the total of all the observations divided by the number of observations. When computing the total, remember to include a salary as many times as it appears. The same is true when ordering the observations to find the median - remember to include a salary as many times as it appears.

Exercise 1.61

KEY CONCEPTS - measures of center, resistant measures

The change of extremes affects the mean, but not the median. To compute the new mean you can figure out the new total (you don't need to add up all the numbers again - just think about how much it has gone up) and divide by the number of observations. Or else you can think about dividing up the salary increases by the number of observations and adding this value to the old mean.

Exercise 1.65

KEY CONCEPTS - standard deviation

There are two points to remember in getting to the answer - the first is that numbers "further apart" from each other tend to have higher variability than numbers closer together. The other is that repeats are allowed. There are several choices for the answer to (a) but only one for (b).

Exercise 1.73

KEY CONCEPTS - linear transformations

This is an exercise in recognizing when a new measurement can be expressed in terms of an old measurement by the equation $x_{new} = a + bx$. This form of transforming an old measurement to a new measurement is called a linear transformation. In part (a) you want to convert water temperature to a new measurement which is the difference between the water temperature and the "normal" body temperature. If the water temperature was 90 degrees, the difference between the water temperature and normal body temperature would be -8.6 degrees (Note that a negative sign would occur whenever the pool temperature was below normal body temperature). We obtained this result by taking x - 98.6, where x is the water temperature. This corresponds to a linear transformation with $a = 1$ and $b = $ -98.6.

Try and set up part (b) yourself. Linear transformations are important in statistics and will appear at several points in the book.

COMPLETE SOLUTIONS

Exercise 1.49
a)

```
   Women  |     | Men
          |  7  | 05
          |  8  | 8
          |  9  | 12
      931 | 10  | 489
        5 | 11  | 3455
      966 | 12  | 6
       77 | 13  | 2
       80 | 14  | 06
      442 | 15  | 1
       55 | 16  | 9
        8 | 17  |
          | 18  | 07
          | 19  |
        0 | 20  |
```

b) For the men, the sum of the 20 scores is 2425 and the mean is $\bar{x} = 121.25$. For the women, the sum of the 18 scores is 2539 and the mean is $\bar{x} = 141.06$. Since the number of scores is even for both groups, the median is the average of the two middle scores. For the men M = 114.5 and for the women M = 138.5. For the women there is really little skewness - the mean exceeds the median because of the outlier, and only slightly. For the men, the distribution is right-skewed, and the mean exceeds the median because of this.

c)

	Men	Women
Minimum	70	101
Q_1	98	126
M	114.5	138.5
Q_3	143	154
Maximum	187	200

The IQR for women is 154 - 126 = 28, and 1.5 × IQR = 37. If we add 37 to the third quartile we get 154 + 37 = 191. Since 200 exceeds this, it is flagged as an outlier.

d) The boxplot makes the comparison of the plots easier. The symmetry of the women's scores, with the exception of the outlier is fairly obvious, as well as the right skewness of the men's distribution. It is also somewhat clearer from examining the boxplot that the women's scores tend to be higher than the men's, while the men's are more variable.

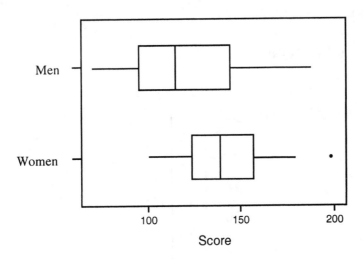

Exercise 1.59

The number of observations (individuals) is 5 + 2 + 1 = 8 and the total of the salaries is

$$(5 \times 25,000) + (2 \times \$60,000) + (1 \times \$255,000) = \$500,000.$$

The mean is $500,000 / 8 = $62,500. Everyone earns less than the mean, except for the owner.
Since there are eight observations, the median is the average of the fourth and fifth smallest observations, which is $25,000. To see this, the eight ordered observations are

$25,000, $25,000, $25,000, $25,000, $25,000, $60,000, $60,000, $255,000

Exercise 1.61

The owner has an increase in salary from $255,000 to $455,000, or an increase of $200,000. The total is increased by this amount, from $500,000 to $700,000 and the new mean is $700,000/8 = $87,500. Another way of thinking about it is that the $200,000 increase averaged among the 8 people is $25,000, so the mean must go up $25,000 to $62,500 + $25,000 = $87,500.

The fourth and fifth smallest observations are still the same, so the median is unaffected.

Exercise 1.65

a) The standard deviation is always greater than or equal to zero. The only way it can equal zero is if all the numbers in the data set are the same. Since repeats are allowed, just choose all four numbers the same to make the standard deviation equal to zero. Examples are 1, 1, 1, 1 or 2, 2, 2, 2.

b) To make the standard deviation large, numbers at the extremes should be selected. So you want to put the four numbers at zero or ten. The correct answer is 0, 0, 10, 10. You might have thought 0, 0, 0, 10 or 0, 10, 10, 10 would be just as good, but a computation of the standard deviation of these choices shows that two at either end is the best choice.

c) There are many choices for (a) but only one for (b).

Exercise 1.73

a) The difference between x and 98.6 is x - 98.6 (positive numbers correspond to pool temperatures above body temperature and negative numbers to pool temperatures below body temperature). In general, $x_{new} = (x - 98.6)$. (a = -98.6 and b = 1)

b) A food with 120 milligrams corresponds to 100% of the RDA, 60 milligrams to 50% and 240 milligrams to 200%.

In general, $(\%RDA) = 100 \dfrac{(\text{number of milligrams in food})}{120}$ (a = 0, b = 10/3)

SECTION 1.3

OVERVIEW

This section considers the use of mathematical models to describe the overall pattern of a distribution. A mathematical model is an idealized description of this overall pattern, often represented by a smooth curve. The name given to a mathematical model that summarizes the shape of a histogram is a **density curve**. The density curve is a kind of idealized histogram. The total area under a density curve is one and the area between two numbers represents the proportion of the data that lie between these two numbers. Like a histogram, it can be described by measures of center, such as the **median** (a point such that half the area under the density curve is to the left of the point) and the **mean** μ (the balance point of the density curve if the curve were made of solid material), and measures of spread, such as the **quartiles** and the standard deviation σ.

One of the most commonly used density curves in statistics is the **normal curve** and the distributions they describe are called **normal distributions**. Normal curves are symmetric and bell-shaped. The peak of the curve is located above the mean and median, which are equal since the density curve is symmetric. The standard deviation is the distance from the mean to the change-of-curvature points on either side. It measures how concentrated the area is around this peak. Normal curves follow the **68-95-99.7 rule**, i.e., 68% of the area under a normal curve lies within one standard deviation of the mean (illustrated in the figure below), 95% within two standard deviations of the mean, and 99.7% within three standard deviations of the mean.

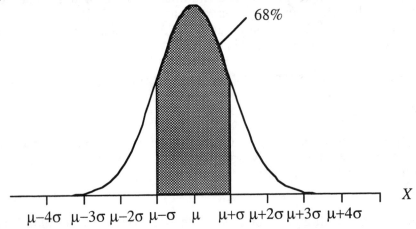

Areas under any normal curve can be found easily if quantities are first **standardized** by subtracting the mean from each value and dividing the result by the standard deviation. This standardized value is sometimes called the z-**score**. If data whose distribution can be described by a normal curve are standardized (all values replaced by their z-scores), the distribution of these standardized values is called the **standard normal** distribution and they are described by the **standard normal curve**. Areas under standard normal curves are easily computed by using a standard normal table such as that found in Table A in the front inside cover of the text.

If we know the distribution of data is described by a normal curve, we can make statements about what values are likely and unlikely, without actually observing the individual values of the data. Although one can examine a histogram or stem-and-leaf plot to see if it is bell-shaped, the preferred method for determining if the distribution of data is described by a normal curve is a **normal quantile plot**. These are easily made using modern statistical computer software. If the distribution of data is described by a normal curve, the normal quantile plot should look like a straight line.

    In general, density curves are useful for describing distributions. Many statistical procedures are based on assumptions about the nature of the density curve that describes the distribution of a set of data. **Density estimation** refers to techniques for finding a density curve that describes a given set of data.

GUIDED SOLUTIONS

Exercise 1.79

KEY CONCEPTS - density curves and area under a density curve

a) In this case, the density curve has unknown height h between 0 and 2, and height 0 elsewhere. Thus, the density curve forms a rectangle with a base of length 2 and a height equal to h. Recall that the area of any rectangle is the product of the length of the base of the rectangle and the height. The area of this density curve is therefore (fill in the blanks)

area = length of base × height = _____ × _____

The total area under a density curve must be 1. What must h be so that this is the case?

Draw a graph of this density curve in the space provided.

              0   2

b) The area of interest is the shaded region in the figure below.

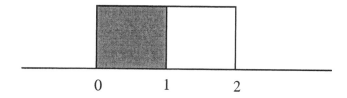

      0   1   2

Compute the area of the shaded region by filling in the blanks below.

area = length of base × height = _____ × _____ = _____

c) Try answering this part on your own, using the same reasoning as in (b). First, shade in the area of interest in the figure below.

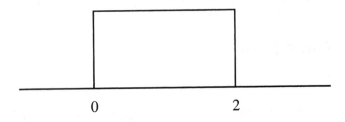

Now compute the area of your shaded region as in (b).

Exercise 1.83

KEY CONCEPTS - the 68-95-99.7 rule for normal curves

Recall that the 68-95-99.7 rule says that for the normal distribution, approximately 68% of the observations fall between the mean minus one standard deviation and the mean plus one standard deviation, 95% of the observations fall between the mean minus two standard deviations and the mean plus two standard deviations, and approximately 99.7% of the observations fall between the mean minus three standard deviations and the mean plus three standard deviations. Also recall that the area under a density curve between two numbers corresponds to the proportion of the data that lies between these two numbers.

In this problem the mean is 336 days and the standard deviation is 3 days. From the 68-95-99.7 rule we have, for example, that 68% of the lengths of all horse pregnancies lie between 336 - (1 × 3) = 333 and 336 + (1 × 3) =339 days.

a) The 68-95-99.7 rule says that approximately 99.7% of the observations fall between the mean minus three standard deviations and the mean plus three standard deviations. The mean minus three standard deviations is 327. This is the lower bound for the shaded region in the figure below. What is the upper bound for the shaded region below? Fill in the space provided in the figure on the next page.

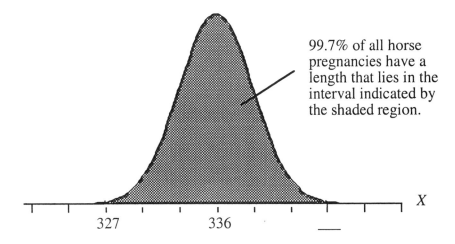

b) We indicated above that the middle 68% of all horse pregnancies have lengths between 333 and 339 days. What percent are either less than 333 or longer than 339? What percent must therefore be longer than 339 (recall that the density curve is symmetric)?

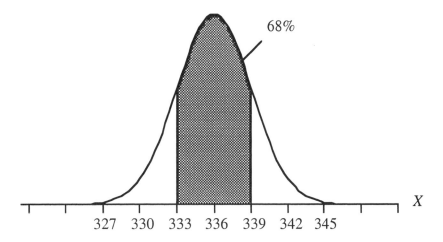

Exercise 1.89

KEY CONCEPTS - computing relative frequencies for a standard normal distribution.

Recall that the proportion of observations from a standard normal distribution that are less than a given value z is equal to the area under the standard normal curve to the left of z. Table A gives these areas. This is illustrated in the figure on the next page.

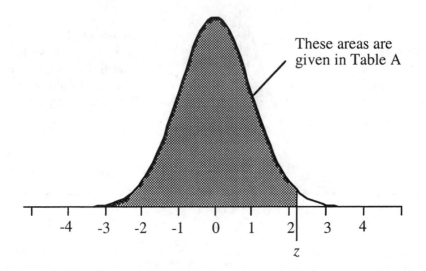

These areas are given in Table A

In answering questions concerning the proportion of observations from a standard normal distribution that satisfy some relation, we find it helpful to first draw a picture of the area under a normal curve corresponding to the relation. We then try to visualize this area as a combination of areas of the form in the figure above, since such areas can be found in Table A. The entries in Table A are then combined to give the area corresponding to the relation of interest.

This approach is illustrated in the solutions that follow.

a) To get you started, we will work through a complete solution. A picture of the desired area is given below.

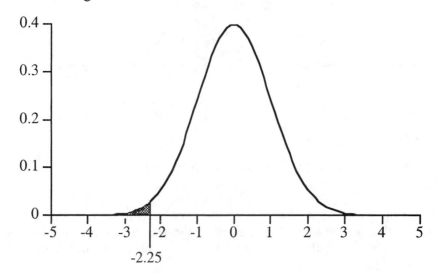

This is exactly the type of area that is given in Table A. We simply find the row labeled -2.2 along the left margin of the table, locate the column labeled .05 across the top of the table, and read the entry in the intersection of this row and column. We find this entry is 0.0122. This is the proportion of observations from a standard normal distribution that satisfies $z < -2.25$.

b) Shade the desired area in the figure below.

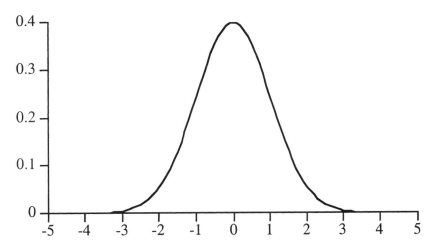

Remembering that the area under the whole curve is 1, how would you modify your answer from part (a)?

area =

c) Try solving this part on your own. To begin, draw a picture of a normal curve and shade the region.

Now use the same line of reasoning as in part (b) to determine the area of your shaded region. Remember, you want to try to visualize your shaded region as a combination of areas of the form given in Table A.

d) To test yourself, try this part on your own. It is a bit more complicated than the previous parts, but the same approach will work. Draw a picture and then try and express the desired area as the difference of two regions for which the areas can be found directly in Table A.

Exercise 1.91

KEY CONCEPTS - finding the value z (the quantile) corresponding to a given area under a standard normal curve

The strategy used to solve this type of problem is the "reverse" of that used to solve Exercise 1.89. We again begin by drawing a picture of what we know; we know the area, but not z. For areas corresponding to those given in Table A we have a situation like the following.

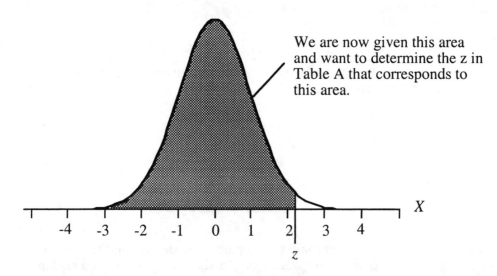

We are now given this area and want to determine the z in Table A that corresponds to this area.

To determine z, we find the given value of the area in the body of Table A (or the entry in Table A closest to the given value of the area). We now look in the left margin of the table and across the top of the table to determine the value of z that corresponds to this area.

If we are given a more complicated area, we draw a picture and then determine from properties of the normal curve the area to the left of z. We then determine z as described above. The approach is illustrated in the solutions below.

a) A picture of what we know is given below. Note that since the area given is larger than 0.5, we know z must be to the right of 0 (recall that the area to the left of 0 under a standard normal curve is 0.5).

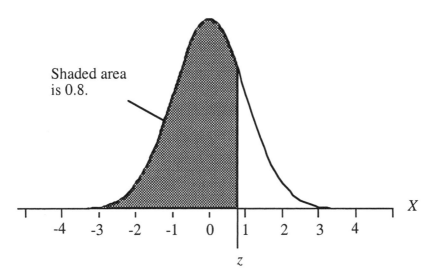

We now turn to Table A and find the entry closest to 0.80. This entry is 0.7995. Locating the z values in the left margin and top column corresponding to this entry, we see that the z that would give this area is 0.84.

b) Try this part on your own. Begin by sketching a normal curve and the area you are given on the curve. On which side of zero should z be located? Thinking about which side of zero a point lies on is a good way to make sure your answer makes sense.

Exercise 1.93

KEY CONCEPTS - computing the area under an arbitrary normal curve

For these problems, we must first state the problem, then convert the question into one about a standard normal. This involves standardizing the numerical conditions by subtracting the mean and dividing the result by the standard deviation. We then draw a picture of the desired area corresponding to these standardized conditions and compute the area as we did for the standard normal, using Table A. This approach is illustrated in the solutions below.

a) *State the problem.* Call the cholesterol level of a randomly chosen young woman X. The variable X has the $N(185, 39)$ distribution. We want the percent of young women with $X > 240$.
    Standardize. We need to standardize the condition $X > 240$. We replace X by Z (we use Z to represent the standardized version of X) and standardize 240. Since we are told that the mean and standard deviation of cholesterol levels are 185 and 39, respectively, the standardized value (z-score) of 240 is (rounded to two decimal places)

$$z\text{-score of } 240 = (240 - 185)/39 = 1.41$$

In terms of a standard normal Z, the condition is $Z > 1.41$.
    A picture of the desired area is

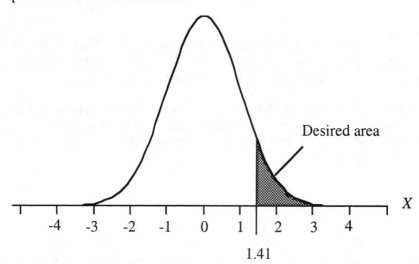

Use the table. The desired area is not of the form given in Table A. However we note that the unshaded area to the left of 1.41 is of the form given in Table A and this area is 0.9207. Hence

shaded area   = total area under normal curve - unshaded area
                 = 1 - 0.9207 = 0.0793

Thus the percent of young women whose cholesterol level X satisfies $X > 240$ is $0.0793 \times 100\% = 7.93\%$.

b) Try this part on your own. First *state the problem*.

Next *standardize*. To do so, compute z-scores to convert the problem to a statement involving standardized values. In terms of z-scores, the condition of interest is

Standardized condition:

Sketch the standard normal curve below and shade the desired region on your curve.

Now *use the table*. Use Table A to compute the desired area. This will be the answer to the question.

Exercise 1.105

KEY CONCEPTS - finding the value x (the quantile) corresponding to a given area under an arbitrary normal curve

To solve this problem, we must use a reverse approach to that used in Exercise 1.93. First we *state the problem*. To make use of Table A, we need to state the problem in terms of areas to the left of some value. Next, we *use the table*. To do so, we think of having standardized the problem and we then find the value z in the table for the standard normal distribution that satisfies the stated condition, i.e., has the desired area to the left of it. We next must *unstandardize* this z value by multiplying by the standard deviation and then adding the mean to the result. This unstandardized value x is the desired result. We illustrate this strategy in the solutions below.

State the problem. We are told in Exercise 1.104 that the WISC scores are normally distributed with $\mu = 100$ and $\sigma = 15$. We want to find the score x that will place a child in the top 5% of the population. This means that 95% of the population scores less than x. We will need to find the corresponding value z for the standard normal. This is illustrated in the figure below.

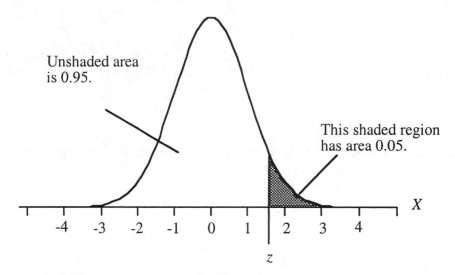

Use the table. The value z must have the property that the area to the left of it is 0.95. Areas to the left are the types of areas reported in Table A. Find the entry in the body of Table A that has a value closest to 0.95. This entry is 0.9495. The value of z that yields this area is seen, from Table A, to be 1.64.

Unstandardize. We now must unstandardize z. The unstandardized value is

$x = $ (standard deviation) $\times z + $ mean $= 15z + 100 = 15 \times 1.64 + 100 = 124.6$
Thus a child must score at least 124.6 to be in the top 5%. Assuming fractional scores are not possible, a child would have to score at least $x = 125$ to place in the top 5%.

Now see if you can determine the score needed for a child to place in the top 1%. Use the same line of reasoning as above.

State the problem. You may find it helpful to draw the region representing the z value corresponding to the top 1%.

Use the table.

Unstandardize.

Exercise 1.111

KEY CONCEPTS - normal quantile plots and determining whether the distribution of a set of data can be described by a normal curve.

The data are given in Exercise 1.27. To make a normal quantile plot you should use statistical software. Consult the user manual for the procedure for your software packages. The first step is to enter the data values. If your software package uses a spreadsheet for data entry, enter the data in a single column. If you have access to an ASCII (text) file containing the data, you should import it into your software package. Then use the appropriate command for making a normal quantile plot. Below is such a plot. Yours should look similar.

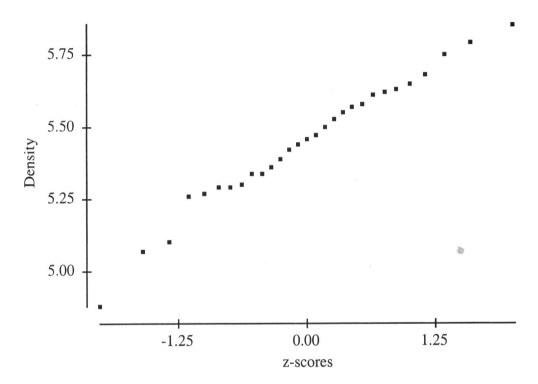

To interpret the plot ask yourself the following questions:

- Do the points appear to follow a straight line?
- If not, in what ways do they deviate? Are there outliers? Are there any unusual "bends" at either end of the plot? Is there evidence of skewness?

Refer to Figures 1.31 to 1.34 in your text for some guidance in interpreting your plot. You might also make a histogram of the data to check your interpretation. Write down your interpretation and check your answer with the solution provided.

COMPLETE SOLUTIONS

Exercise 1.69

a) The area of the density curve in this case is

area = length of base × height = _____ 2 _____ × _____ h _____

In order for this area to be 1, h must be 0.5.
      A graph of the density curve is the following.

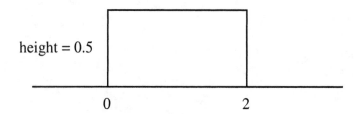

b) The area of interest is the shaded region indicated below.

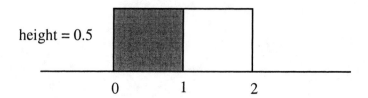

This shaded rectangular region has area = length of base × height = 0.5 × 1 =
0.50.

c) The area of interest is the shaded region indicated below.

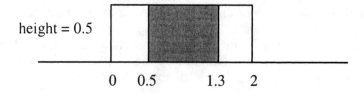

This rectangular region has area = length of base × height = 0.8 × 0.5 = 0.4.

Exercise 1.83

a) The shaded region lies between 327 and 345 days.

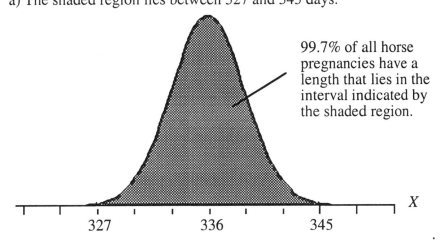

99.7% of all horse pregnancies have a length that lies in the interval indicated by the shaded region.

b) Refer to the figure in the Guided Solutions. If the shaded region gives the middle 68% of the area, then the two unshaded regions must account for the remaining 32%. Since the normal curve is symmetric, each of the two unshaded regions must have the same area and each must account for half of the 32%. Hence each of the unshaded regions accounts for 16% of the area. The rightmost of these regions accounts for the longest 16% of all pregnancies. We conclude that 16% of all horse pregnancies are longer than 339 days.

Exercise 1.89

a) A complete solution was provided in the Guided Solutions.

b) The desired area is indicated below.

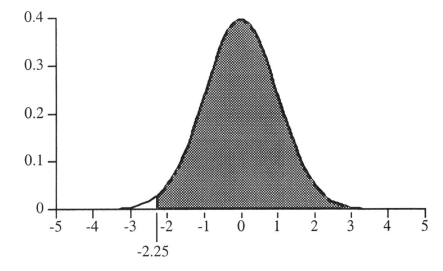

This is not of the form for which Table A can be used directly. However, the unshaded area to the left of -2.25 is of the form needed for Table A. In fact, we found the area of the unshaded portion in part (a). We notice that the shaded area can be visualized as what is left after deleting the unshaded area from the total area under the normal curve.

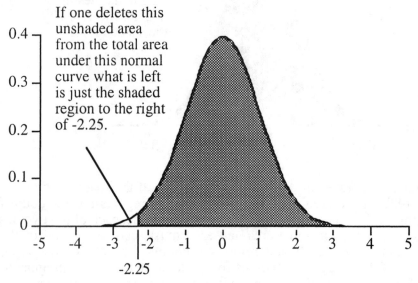

Since the total area under a normal curve is 1, we have

shaded area = total area under normal curve - area of unshaded portion
= 1 - 0.0122 = 0.9878.

Thus the desired proportion is 0.9878.

c) The desired area is indicated in the figure below.

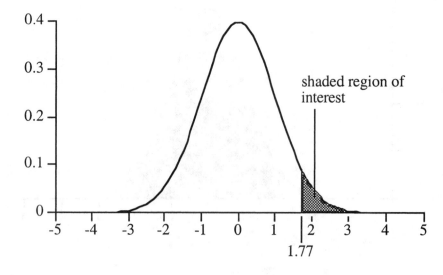

This is just like part (b). The unshaded area to the left of 1.77 can be found in Table A and is 0.9616. Thus

shaded area   = total area under normal curve - area of unshaded portion
              = 1 - 0.9616 = 0.0384.

This is the desired proportion.

d) We begin with a picture of the desired area.

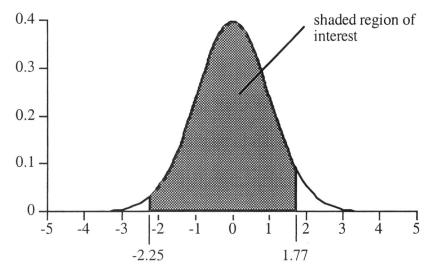

The shaded region is a bit more complicated than in the previous parts, however the same strategy still works. We note that the shaded region is obtained by removing the area to the left of -2.25 from all the area to the left of 1.77.

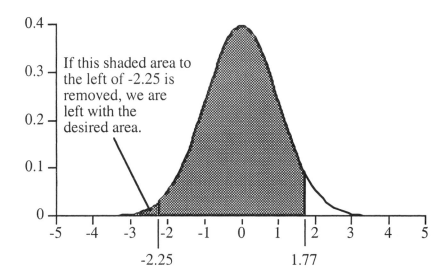

The area to the left of -2.25 is found in Table A to be 0.0122. The area to the left of 1.77 is found in Table A to be 0.9616. The shaded area is thus

shaded area = area to left of 1.77 - area to left of -2.25
= 0.9616 - 0.0122
= 0.9494.

This is the desired proportion.

Exercise 1.91

a) A complete solution was given in the Guided Solutions.

b) A picture of what we know is given below. Note that since the area to the right of 0 under a standard normal curve is 0.5, we know that z must be located to the right of 0.

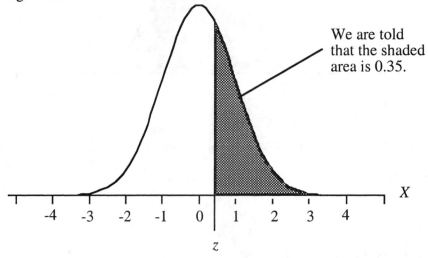

We are told that the shaded area is 0.35.

The shaded area is not of the form used in Table A. However, we note that the unshaded area to the left of z is of the correct form. Since the total area under a normal curve is 1, this unshaded area must be 1 - 0.35 = 0.65. Hence z has the property that the area to the left of z must be 0.65. We locate the entry in Table A closest to 0.65. This entry is 0.6517. The z corresponding to this entry is 0.39.

Exercise 1.93

a) A complete solution was given in the Guided Solutions.

b) *State the problem.* The problem is to find the percent of young women with $200 < X < 240$.

Standardize. We need to first standardize the condition $200 < X < 240$. We replace X by Z (we use Z to represent the standardized version of X) and standardize 200 and 240. Since we are told that the mean and standard deviation are 185 and 39, respectively, the standardized values (z-scores) of 200 and 240 are (rounded to two decimal places).

$$z\text{-score of } 200 = (200 - 185)/39 = 0.38$$

$$z\text{-score of } 240 = (240 - 185)/39 = 1.41$$

Our condition in "standardized" form is $0.38 < Z < 1.41$. A picture of the desired area is

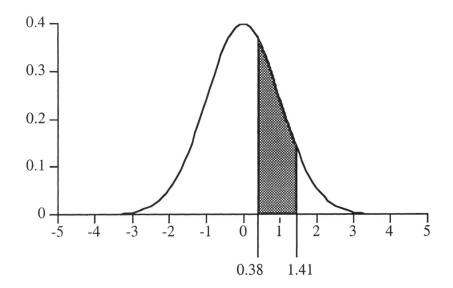

Use the table. This area can be found by determining the area to the left of 1.41, the area to the left of 0.38, and then subtracting the area to the left of 0.38 from the area to the left of 1.41. From Table A

$$\text{area to the left of } 1.41 = 0.9207$$

$$\text{area to the left of } 0.38 = 0.6480$$

and the desired difference is $0.9207 - 0.6480 = 0.2727$. Thus, the percent of young women whose cholesterol level X satisfies $200 < X < 240$ is $0.2727 \times 100\% = 27.27\%$.

Exercise 1.105

The complete solution for the top 5% is given in the Guided Solutions.

For the top 1% we proceed as follows.

State the problem. We want to find the score that will place a child in the top 1% of the population. We first find the corresponding value z for the standard normal. This value z must have the property that the area to the right of it under the standard normal curve is 0.01. This is illustrated in the figure below.

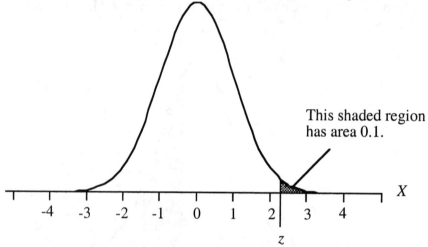

This shaded region has area 0.1.

From the figure we see that the area to the left of z (the unshaded area) is

unshaded area = total area under normal curve - shaded area = 1 - 0.01 = 0.99.

Hence we must find the value z such that the area to the left of it is 0.99.

Use the table. We find the entry in the body of Table A that has value closest to 0.99. This entry is 0.9901. The value of z that yields this area is seen, from Table A, to be 2.33.

Unstandardize. We now must unstandardize z. The "unstandardized" value is

x = (standard deviation) $\times z$ + mean = $15z + 100 = 15 \times 2.33 + 100 = 134.95$

Thus, a child must score at least 134.95 to be in the top 1%. Assuming fractional scores are not possible, a child would have to score at least $x = 135$ to place in the top 1%.

Exercise 1.111

If the data follow a normal distribution, the plot should look approximately like a straight line. In our plot, the three points in the lower-left corner lie below the line drawn through the remaining points. This suggests that the data may be slightly left skewed. Ignoring these three points, the rest of the plot is reasonably straight, suggesting that the remaining data are approximately normal. Making a histogram of the data, one sees the slight left skewness.

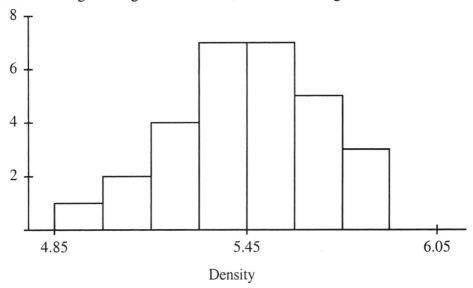

Density

CHAPTER 2

LOOKING AT DATA - RELATIONSHIPS

SECTION 2.1

OVERVIEW

The first chapter provides the tools to explore several types of variables one by one, but in most instances the data of interest are a collection of variables that may exhibit some kind of relationships among themselves. Typically, these relationships are more interesting than the behavior of the variables individually. If we think that one of the variables, x, may explain or even cause changes in another variable, y, we call x an **explanatory variable** and y a **response variable**.

The first tool we consider for examining the relationship between variables is the **scatterplot**. Scatterplots show us two quantitative variables at a time, such as the weight of a car and its MPG (miles per gallon). Using colors or different symbols, we can add information to the plot about a third variable which is categorical in nature. For example, if in our plot we wanted to distinguish between cars with manual or automatic transmissions, we might use a circle to plot the cars with manual transmissions and a cross to plot the cars with automatic transmissions.

When drawing a scatterplot, we need to pick one variable to be on the horizontal axis and the other to be on the vertical axis. When there is a response variable and an explanatory variable, the explanatory variable is always placed on the horizontal axis. In cases where there is no explanatory-response variable distinction, either variable can go on the horizontal axis. After drawing the scatterplot by hand or using a computer, the scatterplot should be examined for an **overall pattern** which may tell us about any relationship between the variables and for **deviations** from it. You should be looking for the **direction**, **form**, and **strength** of the overall pattern. In terms of direction, **positive association** occurs when the variables both take on high values together, while **negative association** occurs if one variable takes high values when the other takes on low values. In many cases, when an association is present, the variables appear to have a **linear relationship**. The plotted values seem to cluster around a line. If the line slopes up to the right, the association is positive; and if the line slopes down to the right, the association is negative. As always, look for **outliers**. The outlier may be far away in terms of the horizontal variable, the vertical variable, or far away from the overall pattern of the relationship.

GUIDED SOLUTIONS

Exercise 2.1

KEY CONCEPTS - explanatory and response variables

a) When examining the relationship between two variables, if you hope to show that one variable can be used to explain variation in the other, remember that the response variable measures the outcome of the study, while the explanatory variable explains changes in the response variable. When you just want to explore the relationship between two variables like score on the math and verbal SAT, then the explanatory-response variable distinction is not important.

In this case, it seems reasonable to view the time spent studying as explaining the grade on the exam. Thus, the grade on the exam is the response and the time spent studying is the explanatory variable. Now try the other parts on your own.

b)

c)

d)

e)

Exercise 2.7

KEY CONCEPTS - drawing and interpreting a scatterplot

a) When drawing a scatterplot, we first need to pick one variable (the explanatory variable) to be on the horizontal axis and the other (the response) to be on the vertical axis. In this data set we are interested in the "effect" of drinking moderate amounts of wine on yearly deaths from heart disease. So wine consumption is the explanatory variable and deaths from heart disease is the response. We have drawn the points corresponding to Australia and Austria in the plot below. Although you will generally draw scatterplots on the computer, drawing a small one like this by hand makes sure that you understand what the points represent.

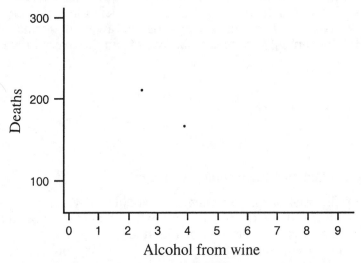

b) We are looking for the form and strength of the relationship. Can the relationship be described with a straight line? Section 2.3 discusses formal methods for drawing a straight line through a set of data, but for now just try to draw a straight line to follow the overall pattern in the scatterplot in part (a) above. Do the points seem to follow the line that you have drawn or are there significant deviations from the pattern? How tight is the scatter about that line?

c) Is the association positive or negative? Do countries with higher wine consumption tend to have higher or lower death rates? You need to be careful with the language you use to describe the relationship. In this example the countries may differ on many other factors besides wine consumption, which may explain the lower death rates due to heart attacks. So avoid the use of expressions such as drinking more wine lowers the risk of heart disease, or wine produces a lower risk of heart disease, as these expressions imply causation. Try and explain in simple language what the data has to say.

Exercise 2.11

KEY CONCEPTS - drawing and interpreting a scatterplot, adding a categorical variable to a scatterplot

a) When drawing a scatterplot, we first need to pick one variable (the explanatory variable) to be on the horizontal axis and the other (the response) to be on the vertical axis. In this data set we are interested in the "effect" of lean body mass on metabolic rate. So lean body mass is the explanatory variable and metabolic rate is the response variable in the plot in the following figure. Although you will generally draw scatterplots on the computer, drawing a small one like this by hand makes sure that you understand what the points represent.

The scatterplot with the points for the females is given below. Add the data for the males to this plot using a different color or plotting symbol.

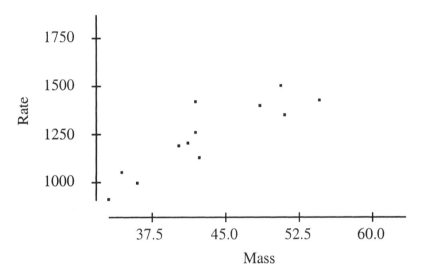

b) Here are some guidelines for examining scatterplots: Do the data show any association? **Positive association** is when the variables both take on high values together. **Negative association** is when one variable takes high values and the other takes on low values. If the plotted values seem to form a line, the variables may have a **linear relationship**. If the line slopes up to the right, the association is positive. If the line seems to slope down to the right, the association is negative. Are there any **clusters** of data? Clusters are distinct groups of observations. As always, look for outliers. An **outlier** may be far away in terms of the horizontal variable or the vertical variable, or far away from the overall pattern of the relationship.

For our plot, is the overall association positive or negative? What is the overall form of the relationship? How strong is the overall relationship?

Is the pattern of the relationship for the men similar to that for the female subjects? If not, how do the male subjects as a group differ from the female subjects as a group?

Exercise 2.16

KEY CONCEPTS - categorical variables in scatter plots

a) In the scatterplot, there should be four observations above each of the levels of nematode count. After adding these points to the graph below, compute the mean of the four observations at each level of nematode count, and put each mean on the graph. Then connect the four means.

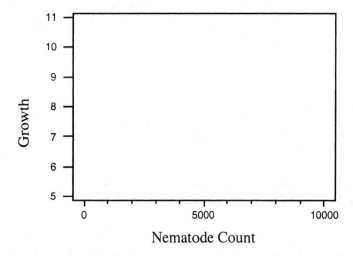

b) What sorts of changes do you see in the means as the nematode count increases?

COMPLETE SOLUTIONS

Exercise 2.1

a) A complete solution was provided in the Guided Solutions.

b) We would probably simply want to explore the relationship between weight and height.

c) We would probably view inches of rain as explaining the yield of corn. Thus, the response is the yield of corn in bushels and the explanatory variable is inches of rain in the growing season.

d) We would probably simply want to explore the relationship between a student's scores on the SAT math exam and scores on the SAT verbal exam.

e) We would probably view a family's income as explaining the years of education their eldest child receives. Thus, the response is the years of education that the eldest child receives and the explanatory variable is the family's income.

Exercise 2.7

a), b)

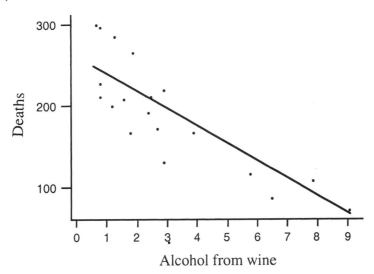

A line does not do a bad job of describing the general pattern. The relationship is moderately strong. In Section 2.2 we will give a numerical measure which describes the strength of the linear relationship.

c) There is a negative association between wine consumption and deaths from heart disease. Those countries in which wine consumption is higher, tend to have a lower rate of deaths from heart disease. (Note: there is nothing in this language which implies causation.)

Exercise 2.11

a) We add the men to the plot. Men are indicated by the x's in the plot on the next page.

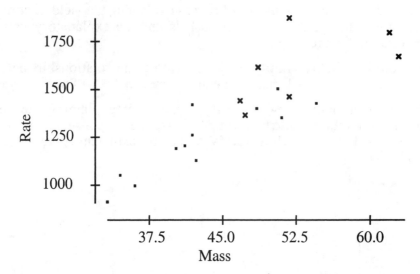

b) As lean body mass increases, or as you move from left to right across the horizontal axis in the scatterplot, the points in the plot tend to rise. This indicates that the association between the variables is positive. The form of the relationship appears to be linear since a straight line seems to be a reasonable approximation to the overall trend in the plot. The relationship is not perfect, but it appears to be moderately strong.

The pattern of the relationship is roughly the same for men and women. The strength of the relationship for females appears to be slightly stronger than for males. The most striking difference between the points corresponding to male and female subjects is that the men are clustered in the upper right of the plot. This is not surprising, since men tend to be larger than women.

Exercise 2.16

a) In the graph below, the circles correspond to the observations and the pluses to the means at each level of nematode count.

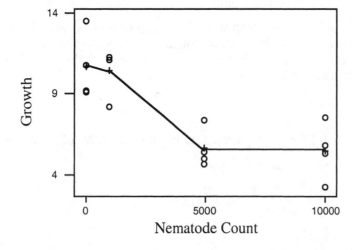

b) The level of growth may decrease slightly when going from 0 to a 1000 count, but then it drops off considerably by 5000 and seems to stay at that level until 10,000. In making these statements we are making some assumptions about the average growth between the levels of nematode counts in our experiments (interpolating). We claimed the growth level wasn't changing between 5000 and 10,000 but we have no data between those values. More problematic is when the drop in level occurs to get from the level at 1000 to the level at 5000. It would be helpful to have had an observation at 3000, as the data suggests there was a fairly sharp decrease in growth at some nematode count between 1000 and 5000, but we can't say more.

SECTION 2.2

OVERVIEW

Scatterplots provide a visual tool for looking at the relationship between two variables. Unfortunately our eyes are not good tools for judging the strength of the relationship. Changes in the scale or the amount of white space in the graph can easily affect our judgment as to the strength of the relationship. **Correlation** is a numerical measure we will use to show the strength of **linear association**.

The correlation can be calculated using the formula

$$r = \frac{1}{n-1} \sum (\frac{x_i - \bar{x}}{s_x})(\frac{y_i - \bar{y}}{s_y})$$

where $\bar{x}$ and $\bar{y}$ are the respective means for the two variables X and Y, and s_x and s_y are their respective standard deviations. In practice, you will probably compute the value of r using computer software or a calculator that finds r from entering the values of the x's and y's. When computing a correlation coefficient there is no need to distinguish between the explanatory and response variables, even in cases where this distinction exists. The value of r will not change if we switch x and y.

When r is positive it means that there is a positive linear association between the variables and when it is negative there is a negative linear association. The value of r is always between 1 and -1. Values close to 1 or -1 show a strong association while values near 0 show a weak association. As with means and standard deviations, the value of r is strongly affected by outliers. Their presence can make the correlation much different than what it might be with the outlier removed. Finally, remember that the correlation is a measure of straight line association. There are many other types of association between two variables, but these patterns will not be captured by the correlation coefficient.

GUIDED SOLUTIONS

Exercise 2.20

KEY CONCEPTS - interpreting and computing the correlation coefficient

a)

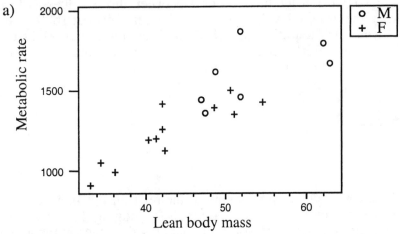

Should the sign of the correlation coefficient be the same for men and women? Is either relationship "stronger?" Are there outliers in either group that might raise or lower the value of the correlation coefficient?

b) Try and use a computer package or a calculator to compute the value of the correlation coefficient. If you do not have access to a calculator or computer package, the required "hand" computations are illustrated below for the men.

$$\bar{x} = 53.10 \qquad\qquad s_x = 6.69$$

$$\bar{y} = 1600.00 \qquad\qquad s_y = 189.2$$

We summarize the calculations for the correlation r in the following table

x	$\dfrac{x - \bar{x}}{s_x}$	y	$\left(\dfrac{y - \bar{y}}{s_y}\right)$	$\left(\dfrac{x - \bar{x}}{s_x}\right)\left(\dfrac{y - \bar{y}}{s_y}\right)$
62.0	1.33034	1792	1.01480	1.35003
62.9	1.46487	1666	0.34884	0.51100
47.4	-0.85202	1362	-1.25793	1.07178
48.7	-0.65770	1614	0.07400	-0.04867
51.9	-0.17937	1460	-0.73996	0.13273
51.9	-0.17937	1867	1.41121	-0.25313
46.9	-0.92676	1439	-0.85095	0.78862

The sum of the values in the last column above is 3.5524. Thus the correlation is

$$r = 3.5524/6 = 0.592 \text{ for the men.}$$

Now you need to either repeat the above "hand" calculation for the 12 women, or learn how to do the calculation on a computer package or calculator.

Women's correlation coefficient =

c)   Mean body mass for men =

Mean body mass for women =

To determine whether the fact that men are heavier on average than women influences the correlation, ask yourself the following. Is the relationship between lean body mass and metabolic rate the same for all weights between 40 and 65 kilograms (the range of all the data)? If the relationship is different for heavier people than lighter people, then the fact that men are heavier could effect the value of the correlation since then men and women might have a different relationship between body mass and metabolic rate (with possibly differing strengths as measured by the correlation coefficient).

d) Is this a linear transformation? What is the effect of a linear transformation on the correlation coefficient?

Exercise 2.25

KEY CONCEPTS - computing the correlation coefficient, the effect of outliers

a) Make your scatterplot in the axes provided.

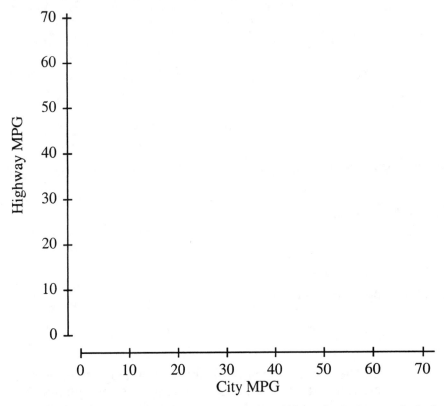

Does the Insight extend the linear pattern of the other cars, or is it far from the line they form?

b) Compute the correlations and enter the results in the space provided. Use statistical software if available. If you are computing the correlations by hand, you may find it useful to organize your calculations as we did in Exercise 2.20.

Correlation with all observations   =

Correlation without the Insight   =

Explain the difference in the two values based on your answer to (a).

Exercise 2.29

KEY CONCEPTS - effect of transformations on the correlation coefficient

a) The first plot is speed (km / h) vs. fuel (liters / 100 km) and the second plot is speed (miles / hour) vs. fuel (gallons / mile). Both are virtually identical except for the labeling of the axes which corresponds to the units that the variables are measured in.

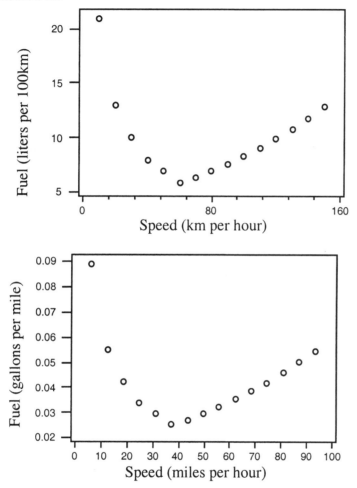

The transformation to change kilometers to miles is miles = kilometers /1.609 and is a linear transformation. What's the transformation to change fuel used in liters/100km to fuel used in gallons/mile? Is it a linear transformation? What is the effect of these two transformations on the numerical value of the correlation coefficient? (**Note:** Since the relationship between speed and fuel consumption is not linear, do you think the correlation coefficient is a good summary of the "strength" of the relationship?)

b)

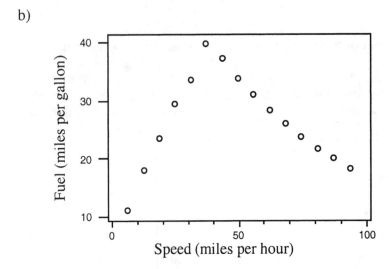

What is the transformation to change fuel used in gallons/mile to miles per gallon? What is the transformation to change liters /100 km to miles per gallon? Is this transformation linear? What is the effect on the correlation coefficient?

Note: The relationship is still not linear so the correlation coefficient is again not a very good summary measure.

Exercise 2.33

KEY CONCEPTS - interpreting the correlation coefficient

The problem is that a correlation close to zero and the quote "good researchers tend to be poor teachers, and vice versa" are not the same. What does a correlation close to zero mean? What would be true about the correlation if "good researchers tend to be poor teachers, and vice versa"?

COMPLETE SOLUTIONS

Exercise 2.20

a) The relationship for the women seems to be tighter around a line. The men's observation of mass = 51.9 and rate = 1867 lowers the value of their correlation.

b)   Men's correlation coefficient = 0.592
       Women's correlation coefficient = 0.876.

c)   Mean body mass for men = 53.10.
       Mean body mass for women = 43.03

The relationship between lean body mass and metabolic rate is roughly a straight line over the range of the data. Thus the fact that men are heavier than women on average would not, of itself, influence the correlation. However, we do notice that the range of values of lean body mass is larger for women than for men. Because the values are more spread out horizontally for women, this may be partly responsible for the larger correlation.

d) Kilograms = 2.2 × Pounds is a linear transformation with $a = 0$ and $b = 2.2$. The value of the correlation is unchanged under linear transformations of the variables.

Exercise 2.25

a)

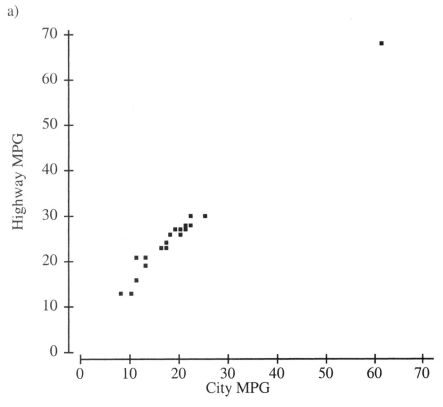

The Insight (outlier in the upper right corner) appears to extend the linear pattern of the other cars.

b) Using statistical software, we obtained the following.

Correlation with all observations   = 0.991
Correlation without the Insight   = 0.956

The correlation is closer to 1 when we include the Insight. As we noted in (a), the Insight appears to extend the linear pattern of the other cars. In so doing, it actually strengthens the visual impression of the linear pattern because it is far from the other points in the plot.

Exercise 2.29

a) The transformation here is $fuel_{(gallons\ per\ mile)} = (1.609/378.5) \times fuel_{(liters\ per\ 100\ km)}$ and is linear. Since the transformation of x and y are both linear, the correlation coefficient is the same for the original and transformed variables. The value of the correlation coefficient is -0.172. Since the relationship is clearly nonlinear, and the correlation only measures the strength of a linear relationship, it would be a mistake to interpret this as a negative relationship between fuel consumption and speed.

b) To go from gallons per mile to miles per gallon we need to take the reciprocal miles per gallon = 1/gallons per mile, or

$$mpg = 378.5 / (1.609 \times fuel_{(liters\ per\ 100\ km)}).$$

This is not a linear transformation, so the numerical value of the correlation coefficient will change. The value of the correlation coefficient is now -.043. A value near zero would suggest a weak relationship but that refers to a weak **linear** relationship. There is clearly a very strong relationship between speed and miles per gallon, albeit nonlinear.

Exercise 2.33

If the correlation were close to zero, there would be no particular linear relationship. Good researchers would be just as likely as bad researchers to be good or bad teachers. The statement that "good researchers tend to be poor teachers, and vice versa" implies that the correlation is negative, not zero.

SECTION 2.3

OVERVIEW

If a scatterplot shows a linear relationship which is moderately strong as measured by the correlation, we would like to draw a line on the scatterplot to summarize the relationship. In the case where there is a response and an explanatory variable, the **least-squares regression** line often provides a good summary of this relationship. A straight line relating y to x has the form $y = a + bx$ where b is the **slope** of the line and a is the **intercept**. The least squares regression line is the straight line $\hat{y} = a + bx$ which minimizes the sum of the squares of the vertical distances between the line and the observed values y. The formula for the slope of the least squares line is

$$b = r\frac{s_y}{s_x}$$

and for the intercept is $a = \bar{y} - b\bar{x}$, where $\bar{x}$ and $\bar{y}$ are the means of the x and y variables, s_x and s_y are their respective standard deviations and r is the value of the correlation coefficient. Typically, the equation of the least squares regression line is obtained by computer software or a calculator with a regression function.

Regression can be used to predict the value of y for any value of x. Just substitute the value of x into the equation of the least squares regression line to get the predicted value for y. Predicting values of y for x values in the range of those x's we observed is called interpolation and is fine to do. However, be careful about **extrapolation** (using the line for prediction beyond the range of x values covered by the data). Extrapolation may lead to misleading results if the pattern found in the range of the data does not continue outside the range.

Correlation and regression are clearly related as can be seen from the equation for the slope, b. However, the more important connection is how r^2, the square of the correlation coefficient, measures the strength of the regression. r^2 tells us the fraction of the variation in y that is explained by the regression of y on x. The closer r^2 is to 1 the better the regression describes the connection between x and y.

GUIDED SOLUTIONS

Exercise 2.37

KEY CONCEPTS - review of straight lines

a) In order to give the equation of a straight line, $y = a + bx$, the first thing is to figure out which variable will play the role of x and which will be y. In this problem

$x =$

$y =$

The only thing remaining is to determine which piece of information represents the intercept or value of y when $x = 0$, and which represents the slope, how much y increases with each unit increase in x. In this problem

$a =$

$b =$

The equation of the line is then

b) Draw the graph on the axes below. Remember, drawing a straight line only requires that you find two points on the line and connect them. The value at zero is easy, so you just need to pick a second value.

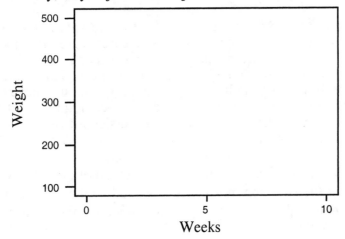

c) When predicting the weight at 2 years, remember the units that age is measured in before doing the calculation. Do you think that a rat's growth pattern will remain the same outside of the range of data? Why or why not?

Exercise 2.45

KEY CONCEPTS - scatterplots, least square regression, interpreting the slope, extrapolation

For purposes of reference, a scatterplot of the world record times (y) against year (x) is given on the next page. The symbol o is used to denote points corresponding to women and x to denote points corresponding to men.

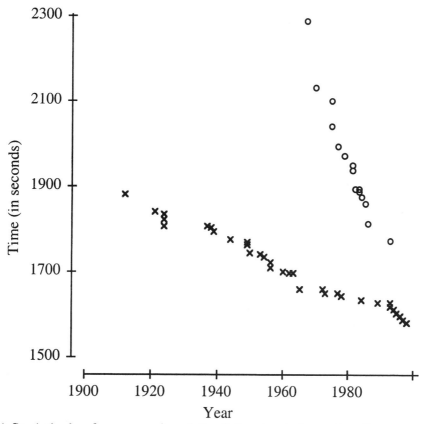

a) Statistical software produced the following information for regressing record time on year.

MEN

	Coefficients	Standard Error	t Stat
Intercept	8167.02	189.40	43.1
Year	-3.29278	0.0966	-34.1

WOMEN

	Coefficients	Standard Error	t Stat
Intercept	41373.2	2717	15.2
Year	-19.9046	1.372	-14.5

Based on this information, give the equation of the least-squares regression lines for predicting record time from year for men and women separately.

MEN

WOMEN

b) What do the slopes tell us about the progress of men and women in the 10,000 meter run? The scatterplot given at the beginning of the problem may help you visualize these slopes.

c) Based on the scatterplot when would you estimate that the women's world record will be the same as the men's?

Exercise 2.51

KEY CONCEPTS - units of measurement for descriptive measures, effect of transformations on summary measures

a) The mean and standard deviation are discussed in Section 2 of Chapter 1 of the text. The correlation is discussed in Section 2 of this chapter. Refer to these sections for help if you have forgotten how to compute the mean, standard deviation, or correlation, or if you have forgotten the units of measurement for these quantities. Compute $\bar{x}$, s_x, $\bar{y}$, s_y and r and then enter the values below. These computations can be done easily using software.

$$\bar{x} = \qquad\qquad s_x =$$

$$\bar{y} = \qquad\qquad s_y =$$

$$r = $$

What are the units of measurement for each of these descriptive measures?

b) You need to figure which of these descriptive measures would have new units of measurement, a new value and what the effect of the linear transformation inches $= (1/2.54) \times$ centimeters has on each quantity.

c) Recall that the slope is $b = r\dfrac{s_y}{s_x}$. Compute this from the quantities you calculated in parts (a) and (b).

$$b = $$

Exercise 2.53

KEY CONCEPTS - r versus r^2

Recall that r^2 tells us the fraction of the variation in y that is explained by the regression of y on x. Is the number 16% the value of r or r^2? If it is r^2, what is the sign of r?

COMPLETE SOLUTIONS

Exercise 2.37

a) The weight is the y variable and age is the x variable. The intercept is the weight at age = 0 or birth and so $a = 100$ grams. The slope is how much the weight goes up with each week and $b = 40$ grams per week. The equation of the line is then $y = 100 + 40x$. Don't forget to include the units in the slope and intercept.

b) At zero the weight is equal to 100 grams, and at 10 weeks the value of weight is $100 + 40(10) = 500$ grams. Just plot these two points on the graph and connect them to get the line.

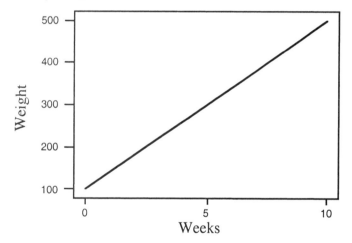

c) This is an example of extrapolation. Rats do not continue to grow at the same rate for two years, just as people wouldn't necessarily grow at the same rate for 20 years. At 2 years or 104 weeks (remember that x is measured in weeks), the predicted weight is $100 + 40(104) = 4260$ grams which would be quite a rat!

Exercise 2.45

a) The equations of the least-squares regression line can be determined from the information provided in the guided solution, or you can determine this from your own statistical software. The entries in the coefficients column give the values of the intercept and slope, respectively. We find

MEN

    Record Time = 8167.02 - 3.29278 × Year

WOMEN

    Record Time = 41,373.2 - 19.9046 × Year

b) Both slopes are negative, indicating that the world record times are going down over time. For men, the decrease in the world record time is 3.29278 seconds per year and for women the decrease is 19.9046 seconds per year *over the range of the data*. The world record times for women have been decreasing at a faster rate than for men over the range of the data.

c) The scatterplot suggests that by approximately the year 2000 men and women will have the same world record times. A more careful mathematical calculation using the equations of the least-squares regression lines shows that the times will be the same in 1999. Of course, the year 2000 has passed and the world record time for women is still greater than for men. So much for extrapolation!

Exercise 2.51

a) Using statistical software we found

$$\bar{x} = 95 \qquad\qquad s_x = 53.3854$$

$$\bar{y} = 12.6611 \qquad\qquad s_y = 8.4967$$

$$r = 0.996$$

The units of measurement are $\bar{x}$ = minutes, $\bar{y}$ = centimeters, s_x = minutes, s_y = centimeters, and r = unitless.

b) The linear transformation inches = (1/2.54)×centimeters changes the mean from $\bar{y}$ in centimeters to (1/2.54)× $\bar{y}$ in inches. Likewise, it changes the standard deviation from s_y in centimeters to (1/2.54)×s_y in inches. The correlation r is unchanged because it is unitless. Thus we get

$$\text{new } \bar{y} \text{ in inches} = 12.6611/2.54 = 4.9847$$

new s_y in inches = 8.4967/2.54 = 3.3452

new r = 0.996.

c) We compute

$$b = r\frac{s_y}{s_x} = 0.996\frac{3.3452}{53.3854} = 0.0624$$

Exercise 2.53

16% is the value of r^2. Hence $r^2 = 0.16$ and $r = \sqrt{0.16} = 0.4$. We take the positive square root because the problem states that, in general, students who attended a higher percentage of their classes earned a higher grade which corresponds to a positive association.

SECTION 2.4

OVERVIEW

Plots of the **residuals**, which are the differences between the observed and predicted values of the response variable, are very useful for examining the fit of a regression line. Features to look out for in a residual plot are unusually large values of the residuals (outliers), nonlinear patterns, and uneven variation about the horizontal line through zero (corresponding to uneven variation about the regression line).

The effects of **lurking variables**, variables other than the explanatory variable which may also affect the response, can often be seen by plotting the residuals versus such variables. Linear or nonlinear trends in such a plot are evidence of a lurking variable. If the time order of the observations is known, it is good practice to plot the residuals versus time order to see if time can be considered a lurking variable.

Influential observations are individual points whose removal would cause a substantial change in the regression line. Influential observations are often outliers in the horizontal direction but they need not have large residuals.

Correlation and regression must be interpreted with caution. Plots of the data, including residual plots, help make sure the relationship is roughly linear and help to detect outliers and influential observations. The presence of lurking variables can make a correlation or regression misleading. *Always remember that association, even strong association, does not imply a cause-and-effect relationship between two variables.*

A correlation based on averages is usually higher than if we had data for individuals. A correlation based on data with a restricted range is often lower than would be the case if we had observed the full range of the variables.

GUIDED SOLUTIONS

Exercise 2.65

KEY CONCEPTS - residuals

a) If you are using statistical software, you should enter the data and use the software to create a scatterplot. Although we are giving you many of the plots, it is a good idea to make sure you understand in each plot why one variable was designated the response and the other the explanatory variable.

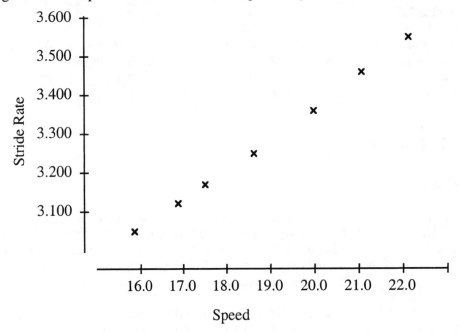

What is the general trend in your scatterplot? Does it appear to be adequately described by a straight line or is curvature present?

b) If you do not have access to a computer or a calculator that will compute the least-squares regression line, you will have to do computations by hand. If you are using statistical software (or a calculator that will compute the equation of the least-squares regression line), you should enter the data and use the software (calculator) to calculate the equation of the least-squares regression line. Write the equation in the space provided below.

c) Recall that the residual for a given speed x is

$$\text{residual = observed stride rate - predicted stride rate}$$
$$\text{predicted stride rate} = a + bx$$

and $a + bx$ is the equation of the least-squares regression line from (b). Statistical software can be used to calculate the residuals directly. If you use statistical software, fill in the values in the residual column only in the table below. If you are calculating the residuals by hand, complete the table below to aid you in systematically calculating the residuals.

speed	observed stride rate	predicted stride rate	residual = observed - predicted
15.86	3.05		
16.88	3.12		
17.50	3.17		
18.62	3.25		
19.97	3.36		
21.06	3.46		
22.11	3.55		

Add the entries in the residual column to verify that the residuals sum to 0 (except for rounding error).

d) If you are using statistical software, the software should allow you to create a plot of the residuals directly. Plot the seven residuals on the axes below.

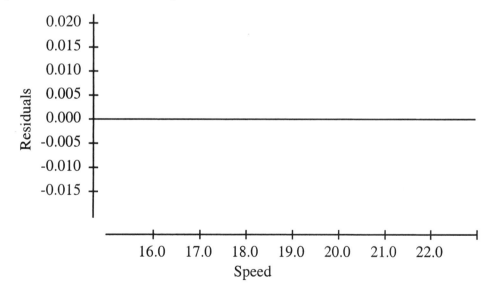

Does it appear that the residuals have a random scatter or is there a pattern present? Do you think that a linear fit is appropriate for these data?

Do any of the points in your plot appear to be influential? Ask yourself if their removal would cause a substantial change in the least-squares line (or if they appear to be outliers in the horizontal direction).

Note: For classes that discussed the topic of DFFITS, you can also calculate the DFFITS to determine if any observations are influential.

Are you provided with any information that would allow you to plot the observations against the time they were made? You might ask yourself, in what form would this information have to be in order to make such a plot (think about what the actual data values represent).

Exercise 2.74

KEY CONCEPTS - regression including a categorical variable, lurking variable

a) A plot of the data is given below. The x corresponds to the right hand times and the circles to the left hand times.

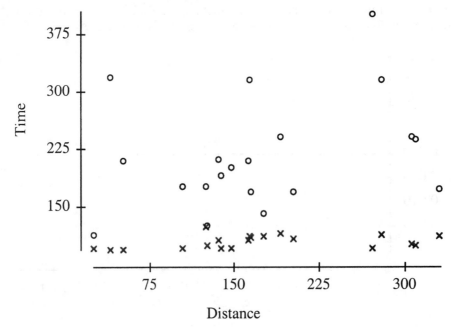

b) In describing the pattern, look for the most striking details. Are there any clear trends for the left-hand observations? The right-hand observations? Are there any differences between the left-hand and right-hand observations?

c) The calculations of the least-squares regression lines of time on distance for each hand are best done using statistical software. Be sure to analyze the left-hand and right-hand observations separately. If you must do them by hand, approach the calculations systematically. Write the two equations of the regression lines in the space provided below.

regression line for left-hand data:

regression line for right-hand data:

Now draw these lines carefully on the scatterplot in (a).

Which line appears to do the better job of predicting time from distance? You should compute the correlation r (or better yet, r^2) associated with each regression line as a possible numerical summary describing the success of the two lines. Can you reconcile the values of your numerical summaries with a visual inspection of the plot?

d) If you are plotting manually, you first calculate the residuals using

$$\text{residual} = \text{time} - [99.364 + 0.028(\text{distance})]$$

for the right-hand observations and

$$\text{residual} = \text{time} - [171.548 + 0.262(\text{distance})]$$

for the left-hand observations. Then you plot your results, remembering that the vertical axis represents the values of the residuals and the horizontal axis time. If you are using statistical software, your plots of the residuals from each regression against the time order of the trials should look similar to those given on the next page. Note the difference in the scales for the vertical axes in the two plots.

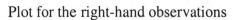

Plot for the right-hand observations

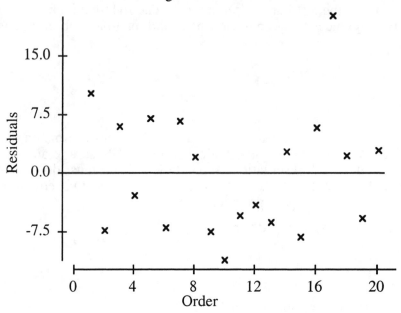

Plot for the left-hand observations

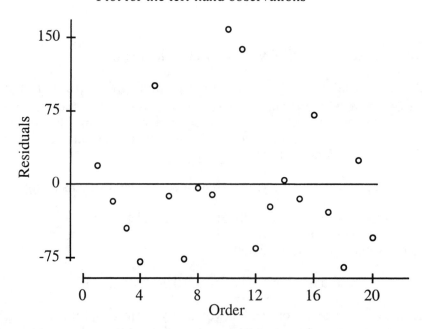

Are any trends or patterns visible in these plots?

Exercise 2.76

KEY CONCEPTS - correlations based on averaged data

Computing the correlation is most easily done using statistical software or a calculator that computes correlation. Regarding whether the correlation would increase or decrease if we had data on the individual stride rates of all 21 runners, note that a correlation based on averages over many individuals is usually higher than the correlation between the same variables based on data for individuals.

COMPLETE SOLUTIONS

Exercise 2.65

a) A plot of the data is given in the Guided Solutions. A straight line "appears" to describe the data quite well.

b) The least-squares regression line

$$\text{Stride Rate} = 1.77 + 0.08(\text{Speed})$$

A scatterplot with this line drawn in is given below.

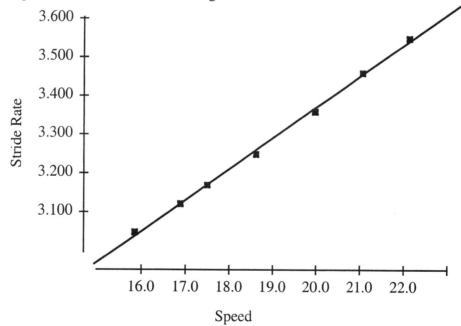

c) For each of the speeds given, we substitute the value into the equation of the least-squares regression line to compute the predicted value of stride rate. The residual is then computed as

residual = observed stride rate - predicted stride rate

For example, for a speed of $x = 15.86$ we compute

predicted stride rate = $1.77 + 0.08(15.86) = 1.77 + 1.27 = 3.04$

and hence

residual = observed stride rate - predicted stride rate = $3.05 - 3.04 = 0.01$.

We summarize the results in the table below (to two decimal places)

speed	observed stride rate	predicted stride rate	residual
15.86	3.05	$1.77 + 0.08(15.86) = 3.04$	0.01
16.88	3.12	$1.77 + 0.08(16.88) = 3.12$	0.00
17.50	3.17	$1.77 + 0.08(17.50) = 3.17$	0.00
18.62	3.25	$1.77 + 0.08(18.62) = 3.26$	-0.01
19.97	3.36	$1.77 + 0.08(19.97) = 3.37$	-0.01
21.06	3.46	$1.77 + 0.08(21.06) = 3.46$	0.00
22.11	3.55	$1.77 + 0.08(22.11) = 3.54$	0.01

While these calculations can be done by hand, they are much more easily obtained using statistical software and should agree (to two decimal places) with the above. If you sum the entries in the residual column, you can easily see they sum to 0.

d) A plot of the residuals against speed is given below

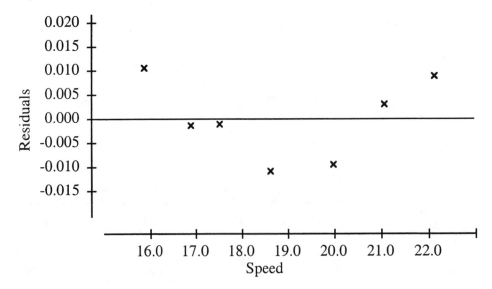

The pattern is curved (U-shaped) and indicates that the linear fit is not completely adequate. Recall that the scatterplot in (a) suggested that the linear fit was quite good. This demonstrates that the residual plot can be more informative than the scatterplot concerning the adequacy of the linear fit.

There do not appear to be any influential observations in the plot. Since we are not told the time at which observations were made we cannot plot the residuals against the time observations were made. Note that since observations are actually the averages for 21 runners, we would have to know that all observations on the 21 runners at a given speed were made at the same time in order to even be able to make such a plot.

Note: For classes that discussed DFFITS, the values of DFFITS are given below. These are most easily computed using statistical software.

Speed	Stride Rate	DFFITS
15.86	3.05	1.6635023
16.88	3.12	-0.08839246
17.50	3.17	-0.05792908
18.62	3.25	-0.59243516
19.97	3.36	-0.55913409
21.06	3.46	0.24486537
22.11	3.55	1.4562468

The first and last entries listed are the largest values of DFFITS and hence the most influential observations. Neither point appears to be particularly influential, however.

Exercise 2.74

a) The plot is given in the Guided Solutions.

b) There does not appear to be any clear relation between distance and time. The most striking feature of the plot is that nearly all the left-hand observations (the o's) lie above the right-hand observations (the x's). This means that left-hand trials had longer times for a given distance than right-hand trials. The subject performed consistently better with the right-hand, suggesting the subject is right-handed.

c) Least-squares regression for the right-hand observations.

$$time = 99.364 + 0.028(distance)$$

Least-squares regression for the left-hand observations.

$$time = 171.548 + 0.262(distance)$$

These lines have been drawn on our scatterplot, reproduced on the next page.

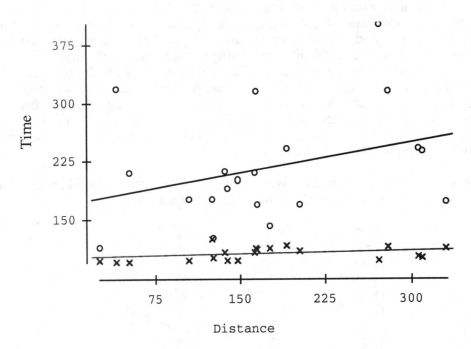

Visually, the regression line for the right-hand data appears to do a better job of predicting time from distance, since the x's appear to be more tightly clustered about this line than the o's are about the regression line for the left-hand data. This would suggest that the value of r^2 would be much higher for the right-hand data. The regression for the right-hand observations has $r^2 = 9.3\%$, i.e., the least-squares regression of time on distance explains 9.3% of the variation in the time values. The regression for the left-hand observations has $r^2 = 10.1\%$, i.e., the least-squares regression of time on distance explains 10.1% of the variation in the time values. Thus the regression for the left-hand observations actually explains a higher percentage of the variation in time than for the right-hand observations.

For the right-hand data there is little variation in times to explain. The times are all quite close to the mean value of 104.25 and this mean is a pretty good predictor of times *without* using distance as an explanatory variable (this is why the least squares regression line is almost horizontal and why r^2 is so low. Distance adds little information to help with predictions). For the left-hand the r^2 value agrees with our visual impression of the plot. There is a lot of variability and distance explains little, resulting in a low r^2.

d) From the equations of the two least-squares regression lines, we can calculate the residuals using the general expression

residual = observed value of time - predicted value of time

For the right-hand observations, this becomes

$$\text{residual} = \text{time} - [99.364 + 0.028(\text{distance})]$$

For the left-hand observations, this becomes

$$\text{residual} = \text{time} - [171.548 + 0.262(\text{distance})]$$

The following tables summarize the results of the above calculations.

Order	residuals(right-hand)	residuals(left-hand)
1	10.2369110	18.503049
2	-7.2858363	-17.829839
3	5.9622312	-44.786498
4	-2.9367652	-79.600335
5	7.0157385	100.708530
6	-7.0176886	-11.607554
7	6.6488317	-76.685980
8	2.0273459	-4.184077
9	-7.5505414	-10.278813
10	-11.049462	158.350190
11	-5.5037426	137.909280
12	-4.0652097	-65.033543
13	-6.3311991	-22.997814
14	2.7628582	3.620668
15	-8.1009221	-14.377606
16	5.7144264	71.165760
17	20.0824860	-28.422228
18	2.2988909	-84.930362
19	-5.8267679	24.920744
20	2.9184153	-54.443579

These residuals can then be plotted against the order in which the observations were taken. The plots are given in the Guided Solutions.

There is no clear pattern in either plot that would suggest the subject got better in later trials due to learning or got worse due to fatigue. Time order does not appear to be a lurking variable.

Exercise 2.76

The value of $r = 0.999$.

This correlation is based on averaged data, namely the average stride rates of 21 runners at each of the seven values of speed (note that such data might have been collected by having the runners run on a treadmill where speed can be controlled). If we had the data on the individual stride rates of all 21 runners, we would expect the correlation to decrease (be less than 0.999).

SECTION 2.5

OVERVIEW

An observed association between two variables can be due to several things. It can be due to a **cause-and-effect** relationship between the variables. It can also be due to the effects of **lurking variables**, i.e., variables not directly studied that may effect the response and possibly the explanatory variable. Lurking variables may operate through **common response**, in which case changes in both the explanatory and response variables are caused by changes in the lurking variable. Lurking variables may also cause **confounding**, in which case both the explanatory variable and the lurking variables cause changes in the response, but we cannot distinguish their individual effects.

The best way to determine if an association is due to a cause-and-effect relationship between the explanatory variable and the response is through an **experiment** in which we control the influences of other variables. In the absence of good experimental evidence, be cautious in accepting claims of causation. Good evidence requires an association that appears consistently in many studies, a clear explanation for the alleged cause-and-effect relationship, and careful examination of possible lurking variables.

GUIDED SOLUTIONS

Exercise 2.82

KEY CONCEPTS - explaining causation, lurking variables

Ask yourself the following questions.

• Was the study an **experiment** in which the influences of other variables were controlled?

• If the study was not an experiment,

> Is there information that the observed association appears consistently in many studies?

> Is there a clear explanation for the alleged cause-and-effect relationship?

> Is there evidence that possible lurking variables have been ruled out as possible causes?

Exercise 2.87

KEY CONCEPTS - lurking variables

Ask yourself, what sorts of students are likely to study foreign language for at least two years. How are such students likely to do in other subjects?

Exercise 2.89

KEY CONCEPTS - evidence for causation in studies that are not experiments

In order to determine what kinds of information you would seek in records, ask yourself

> Would the records contain any information that the observed association appears consistently in many studies or settings?

> Would the records provide a clear explanation for the alleged cause-and-effect relationship?

> Would information in the records allow you to identify or rule out possible lurking variables as causes?

COMPLETE SOLUTIONS

Exercise 2.82

The data are intriguing since they are based on such a large number of operations, but by themselves they do not prove that anesthetic C is causing more deaths than the others. The main weakness of the study is that it was not an experiment in which the influences of other factors were controlled. Some case might still be made for causation, but we would need to know that similar results have been observed in other studies, would need a clear explanation as to why anesthetic C might cause more deaths than the others, and would need evidence that the effects of lurking variables have been ruled out. Unfortunately, no such information is given.

A possible lurking variable that needs to be ruled out is the type of operations for which the anesthetics are used. Is anesthetic C used more often than the others in operations involving long, difficult, and risky surgery? If so, this might explain why the death rate is higher for anesthetic C.

Exercise 2.87

The explanatory variable in this study is whether or not a student has studied a foreign language for at least two years. The response variable is student's score on an English achievement test.

Probably the most important lurking variable is "quality of student" (how hardworking the student is, innate intelligence, innate language ability). Students that are willing to take at least two years of foreign language are also likely to be more serious or talented students that are likely to work hard or do well in other subjects.

Exercise 2.89

The records are not likely to provide us with a clear explanation as to why anesthetic C would have a higher death rate. This probably requires a medical, biological, or chemical explanation.

The records would also allow us to see if the anesthetics are used for different purposes. Perhaps certain anesthetics are used in simple, routine types of surgery, while others are used in complicated, more risky types of surgery. In general, we should look for other patterns of association to identify lurking variables.

The records would also allow us to see if the pattern is in a variety of settings which would allow us to rule out possible lurking variables. We might look to see if the pattern repeats itself if we restrict to cases from specific geographic regions, specific types of hospitals, specific types of operations, or specific types of patients.

SECTION 2.6

OVERVIEW

Many relationships between two quantitative variables are nonlinear rather than linear. Sometimes, nonlinear relationships can be changed into linear relationships by **transforming** one or both variables. **Power transformations** consist of transforming a variable t to the variable t^p and are the most commonly used transformations. It is sometimes convenient to consider the **ladder of power transformations** that corresponds to transforming t to

$$\frac{t^p - 1}{p}$$

The logarithm $\log t$ fits into the ladder of power transformations and corresponds to $p = 0$.

A function $f(t)$ is **monotonic** if it changes in one direction (only increases or only decreases) as t increases. The power transformation is monotonic when the variable we are transforming takes only positive values. In this case there is an inverse transformation that returns the transformed data back to their original values. The effect of power transformations on data becomes stronger as we move farther away from a linear transformation, i.e., as p moves farther away from the value 1.

When we have reason to believe that data are governed by some mathematical model, power transformations are very useful. The **exponential growth model** $y = ab^x$ becomes linear when we plot log y against x. The **power law model** $y = ax^p$ becomes linear when we plot log y against log x.

We can fit exponential growth and power models to data by first transforming the data so that the relationship of the transformed variables looks linear, then fitting the least-squares regression line to the transformed data, and finally doing the inverse transformation.

GUIDED SOLUTIONS

Exercise 2.93

KEY CONCEPTS - exponential growth and power models

a) In this example we would consider Activity to be the response variable (y) and Days to be the explanatory variable (x). A scatterplot of these variables is given below.

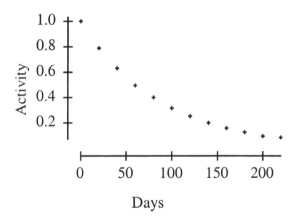

This has a distinct curvature and is nonlinear. To determine if an exponential growth model is more appropriate, we can plot log(Activity) against Days. If the plot is linear with a positive slope, than the exponential growth model provides a good description of the data. If the plot is linear with a negative slope, than the exponential decay model provides a good description of the data. To determine if a power law model is more appropriate, we can plot log(Activity) against log(Days). If the plot is linear, than the power law model provides a good description of the data. Make both plots and determine which model best describes the data. You can either make the plots by hand in the space provided on the next page, or you can use statistical software.

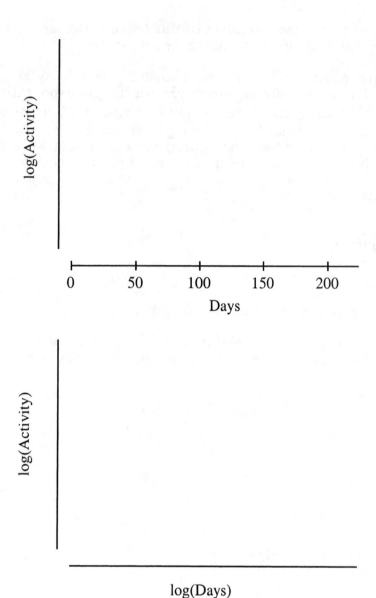

b) Fit a least-squares regression line to the transformed data. What is the equation of the line in terms of the transformed data?

Equation:

Now use the appropriate inverse transformations to express the equation in terms of the original data. Recall that the inverse transformation of $\log(x)$ is $10^{\log(x)}$.

Equation:

c) Is there any feature of the scatterplots you made in (a) that would indicate whether the data might be from actual laboratory experiments or calculated from a theoretical model?

Exercise 2.105

KEY CONCEPTS - transformations, power law

To fit a power law, first compute log(Weight) and log(Lifespan). Then fit a least-squares regression model to the logarithms. Write the equation of your regression model below.

Equation: log(Lifespan) = _____ + _____×log(Weight)

Recall that the slope of this model corresponds to the power p of the power law model Lifespan = a×(Weight)^p. Is the slope approximately 0.2? You may also want to examine a plot of log(Lifespan) against log(Weight) to assess whether a power law model seems reasonable. What do you conclude?

Substitute a weight of 143 into the least-squares regression model to predict log(Lifespan), then raise 10 to this power to get the predicted Lifespan for humans. What do you find?

COMPLETE SOLUTIONS

Exercise 2.93

a) The plots of log(Activity) against Days and log(Activity) against log(Days) are given on the next page.

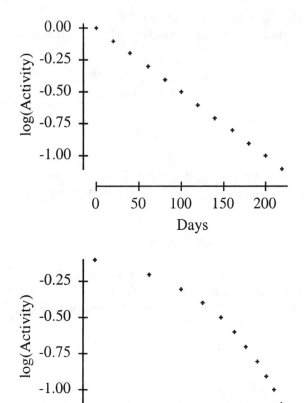

The plot of log(Activity) against Days is a straight line with a negative slope. This suggests that an exponential decay model is a good description of the data.

b) Using statistical software, we found the equation of the least-squares line for log(Activity) as a function of Days to be

Equation: log(Activity) = -0.00002907 - 0.00501799 × Days

We apply the inverse transformation on each side of this equation by raising 10 to power equal to the quantity given on each side. This yields

Equation: Activity = $10^{-0.00002907-0.00501799 \times Days}$

c) The plot of log(Activity) against Days is nearly a perfect straight line. The fit is so good that it is more consistent with data calculated from a theoretical model than data from an actual laboratory experiment. One would expect data from an actual experiment to contain some measurement error and hence not to have such a perfect fit.

Exercise 2.105

Using statistical software we obtain the following equation for the least-squares regression of log(Lifespan) on log(Weight).

Equation: log(Lifespan) = 0.667 + 0.257×log(Weight)

The slope is 0.257, so this yields a power law model with $p = 0.257$. This is reasonably close to 0.2. A plot of log(Lifespan) against log(Weight) is given below.

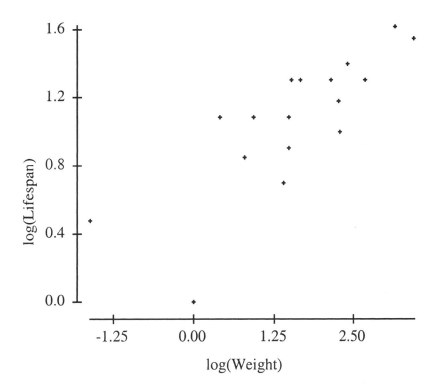

The general pattern is linear, and since the fitted model has a power near 0.2 , a power law model with power 0.2 may not be an unreasonable model.

If we substitute a Weight of 143 into our least-squares regression equation, we would predict

log(Lifespan) = 0.667 + 0.257×log(143) = 1.22.

Thus we would predict

Lifespan = $10^{1.22}$ = 16.6 years.

Obviously humans are an exception to the rule.

CHAPTER 3

PRODUCING DATA

SECTION 3.1

OVERVIEW

Chapters 1 and 2 describe methods for exploring data. Such **exploratory data analysis** is used to determine what the data tell us about the variables measured and their relations to each other. Conclusions apply to the data observed and may not generalize beyond these data.

Statistical inference produces answers to specific questions, along with a statement of how confident we are that the answer is correct. Answers are usually intended to apply beyond the data observed. This requires careful **production of data** appropriate for answering the specific questions asked.

Data can be produced in many ways. **Anecdotal data** based on a few isolated cases is usually unreliable. **Available data** collected for other purposes, such as data produced by government agencies, can be helpful but again is not always reliable. **Sampling** selects a part of a population of interest to represent the whole. Done properly, sampling can yield reliable information about a population. **Observational studies** are investigations in which one simply observes the state of some population, usually with data collected by sampling. Even with proper sampling, data from observational studies are generally not appropriate for investigating cause-and-effect relations between variables. **Experiments** are investigations in which data are generated by active imposition of some treatment on the subjects of the experiment. Properly designed experiments are the best way to investigate cause-and-effect relations between variables.

GUIDED SOLUTIONS

Exercise 3.1

KEY CONCEPTS - anecdotal data

Is this data the result of an experiment? Can there be other explanations for what has occurred?

Exercise 3.5

KEY CONCEPTS - explanatory and response variables, experiments

Remember, experiments are investigations in which data are generated by *active* imposition of some treatment on the subjects. Was that done in this example?

To identify the explanatory and response variables, think about what the experimenter is trying to demonstrate with this study and what is going to be measured.

COMPLETE SOLUTIONS

Exercise 3.1

An isolated, anecdotal case is not a good basis for drawing general conclusions. Although it seems that two aquaintances is a high number to develop brain tumors, we do not have any basis for determining whether this is related to their use of cell phones. Possibly there is another environmental factor related to the incidence of brain tumors that her friends have been exposed to, or there may not be any connection whatsoever between these two incidents of brain tumors.

Exercise 3.5

This is an experiment as a treatment is imposed on the students. The explanatory variable is the teaching method used (standard or computer assisted). The response variable is the increase in knowledge of cell biology as measured by the increase in test score.

SECTION 3.2

OVERVIEW

Experiments are studies in which one or more **treatments** are imposed on experimental **units** or **subjects**. A treatment is a combination of levels of the explanatory variables, called **factors**. The **design** of an experiment is a specification of the treatments to be used and the manner in which units or subjects are assigned to these treatments. The basic features of well-designed experiments are **control**, **randomization**, and **replication**.

Control is used to avoid confounding (mixing up) the effects of treatments with other influences such as lurking variables. One such lurking variable is the **placebo effect**, which is the response of a subject to the fact of receiving any treatment. The simplest form of control is **comparative experimentation** which involve comparisons between two or more treatments. One of these treatments may be a **placebo** (fake treatment), and those subjects receiving the placebo are referred to as a **control group**.

Randomization uses a well-defined chance mechanism to assign subjects to treatments. It is used to create treatment groups which are similar, except for chance variation, prior to application of treatments. Randomized, comparative experiments are used to prevent **bias**, or systematic favoritism of certain outcomes. **Tables of random digits** or computer programs that generate random numbers are well-defined chance mechanisms that are used to carry out randomization. In either case, numerical labels are assigned to experimental units and random numbers from the table or computer software determine which labels (units) are assigned to which treatments.

Replication is the use of many units in an experiment and is used to reduce the effect of any chance variation between treatment groups arising from randomization. Replication increases the sensitivity of an experiment to differences in treatments.

Additional control in an experiment can be achieved by forming experimental units into **blocks** that are similar in some way which is thought to affect the response. In a **block design**, units are first formed into blocks and then randomization is carried out separately in each block. **Matched pairs** are a simple form of blocking used to compare two treatments. In a matched pairs experiment either the same unit (the block) receives both treatments in a random order or very similar units are matched in pairs (the blocks). In the latter case, one member of the pair receives one of the treatments and the other member the remaining treatment. Members of a matched pair are assigned to treatments using randomization.

Good experiments require attention to details. **Double-blind** experiments are ones in which neither the subject nor the person measuring the response is aware of what treatment is being used. **Lack of realism** in an experiment can prevent us from generalizing the results.

GUIDED SOLUTIONS

Exercise 3.11

KEY CONCEPTS - identifying experimental units or subjects, factors, treatments, and response variables

You need to read the description of the study carefully. To identify the experimental units, ask yourself, exactly on what were the experimental conditions applied?

To identify the factors, ask yourself what question did the experiment wish to answer? What variables does the description say the answer depends on? These are the factors.

To identify the treatments, what combinations of values of the factors were actually used in the experiment? These are the treatments.

To identify the response variables, ask yourself what was measured on the subjects after exposure to the treatments? This is the response variable.

Exercise 3.13

KEY CONCEPTS - design of an experiment, randomization

To begin, identify the subjects, the factors, the treatments, and the response variable. Now outline your design. Be sure to specify

• How many treatments are there, hence how many groups of subjects must you form?

• How will you assign subjects to treatment groups?

• What are the treatments, i.e., what will each subject be required to do?

• What response will you measure and how will you decide if the treatments differ in their effect?

You can outline your design in words or with a picture.

The list of names has been reproduced below. Assign a numerical label to each. Be sure to use the same number of digits for each label.

Acosta	Farouk	Liang	Solomon
Asihiro	Fleming	Maldonado	Trujillo
Bennett	George	Marsden	Tullock
Bikalis	Han	Montoya	Valasco
Chen	Howard	O'Brian	Vaughn
Clemente	Hruska	Ogle	Wei
Duncan	Imrani	Padilla	Wilder
Durr	James	Plochman	Willis
Edwards	Kaplan	Rosen	Zhang

Now start reading line 130 in Table B. Read across the row in groups of digits equal to the number of digits you used for your labels (for example, if you used two digits for labels, read line 130 in pairs of digits). You will need to keep reading until you have selected all the names for the first treatment. This may require you to continue on to line 131, line 132, and subsequent lines. After you have selected the names for treatment 1, continue in Table B to assign the nine people to receive treatment 2 and then nine to receive treatment 3. The remaining names are assigned to treatment 4.

Exercise 3.29

KEY CONCEPTS - matched pairs design, randomization

The first thing you should do is identify the subjects, the factor, the treatments, and the response variable. Next, decide what are the matched pairs in this experiment. How will you use a coin flip to assign members of a pair to the treatments? What will you measure and how will you decide whether the right-hand tends to be stronger in right-handed people?

Exercise 3.32

KEY CONCEPTS - block designs

a) To assist you, we have arranged the subjects and their excess weight in order of increasing excess weight. Now decide which are the five blocks. (Due to ties in the excess weights, the choice of blocks may not be unique).

Williams 22	Santiago 27	Brunk 30	Jackson 33	Birnbaum 35
Festinger 24	Mann 28	Obrach 30	Stall 33	Tran 35
Hernandez 25	Smith 29	Rodriguez 30	Brown 34	Nevesky 39
Moses 25	Kendall 30	Loren 32	Dixon 34	Wilansky 42

b) Label the names in each block (you should need only the labels 1, 2, 3, 4 in each block since there are only four people in a block), choose a line in Table B, start reading from left to right, assigning each member of a given block to one of the four treatments.

Write your results below, as indicated.

Line in Table B used =

Subjects on regimen A =

Subjects on regimen B =

Subjects on regimen C =

Subjects on regimen D =

Exercise 3.35

KEY CONCEPTS - properties of random digits

Write your answers (True or False) in the space provided. Remember that a table of random digits is defined to be a list of the digits 0, 1, 2, 3, 4, 5, 6, 7, 8, 9 that has the following properties:

1. The digit in any position in the list has the same chance of being any one of 0, 1 2, 3, 4, 5, 6, 7, 8, 9.

2. The digits in different positions are independent in the sense that the value of one has no influence on the value of any other.

Additional properties are listed in the text below the box containing the definition above.

a) _____

b) _____

c) _____

COMPLETE SOLUTIONS

Exercise 3.11

The experimental units are the households that were called.

There are two factors in this study related to the nature of the introductory remarks. One factor is the information provided about the caller (name only, university being represented only, or name and university being represented). The other factor is whether survey results were offered (yes or no).

The treatments are combinations regarding the information about the caller and whether the survey results were offered. Thus there are six treatments ((1) name only and survey results offered, (2) name only and survey results not offered, (3) university represented only and survey results offered, (4) university represented only and survey results not offered, (5) name and university provided and survey results offered, and (6) name and university provided and survey results not offered).

The response variable is whether or not the interview was completed.

Exercise 3.13

In this case, the subjects are the 36 headache sufferers who have agreed to participate in the study. The two factors are antidepressant (placebo or antidepressant given) and stress mangagement training (given or not given). These form the four treatments which we label as:

Treatment 1: Antidepressant and no stress management training.

Treatment 2: Placebo (no antidepressant) and no stress management training.

Treatment 3: Placebo (no antidepressant) and stress management training.

Treatment 4: Antidepressant and stress management training.

The response variables are number of headaches over the study period and some measure of the severity of these headaches. The problem does not specify how the severity is to be measured.

The study should be done as follows. Subjects should be randomly assigned to treatments, with nine being assigned to treatment 1, nine being assigned to treatment 2, nine being assigned to treatment 3 and the remainder assigned to treatment 4. Each subject follows their treatment regimen over the course of the study. The average number of headaches for each treatment should be calculated and the results for the four groups compared, as well as a comparison of the severity. A picture which summarizes the experimental design is given below.

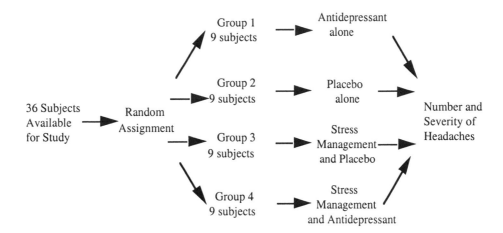

Although you can use the applet or Table B, we illustrate the use of Table B to carry out the random assignment of the subjects to the treatments. First label the 36 names using 2-digit labels. We use the convention of starting with the label 00 and label down the columns. Of course, one could start with another number (such as 01) and label across rows if one wished. The names with labels are

00 Acosta	09 Farouk	18 Liang	27 Solomon
01 Asihiro	10 Fleming	19 Maldonado	28 Trujillo
02 Bennett	11 George	20 Marsden	29 Tullock
03 Bikalis	12 Han	21 Montoya	30 Valasco
04 Chen	13 Howard	22 O'Brian	31 Vaughn
05 Clemente	14 Hruska	23 Ogle	32 Wei
06 Duncan	15 Imrani	24 Padilla	33 Wilder
07 Durr	16 James	25 Plochman	34 Willis
08 Edwards	17 Kaplan	26 Rosen	35 Zhang

Line 130 from Table B is reproduced below. We should read line 130 in pairs of digits from left to right. We have placed vertical bars between consecutive pairs to indicate how we have read the table. We underline those pairs that correspond to labels in our list and that have not been previously selected. On line 130 we have

69|<u>05</u>|1 6|48|<u>17</u>| 87|17|4 0|95|17| 84|53|4 0|64|89| 87|<u>20</u>|1 9|72|45

We only find 5 of our labels so we need to continue reading on line 131.

05|<u>00</u>|7 1|66|<u>32</u>| 81|19|4 1|48|73| <u>04</u>|19|7 8|55|76| 45|19|5 9|65|65

We now have eight labels and continuing to line 132 gives us the 9th label for the group assigned to treatment 1.

68|73|<u>2 5</u>|52|59| 84|29|2 0|87|96| 43|16|5 9|37|39| 31|68|5 9|71|50

The treatment 1 group consists of Clemente, James, Kaplan, Marsden, Maldonado, Acosta, Wei, Chen, and Plochman. Continuing in line 132 (and ignoring pairs corresponding to previously selected subjects) we assign the next nine subjects to treatment 2,

|52|59| 84|<u>29</u>|2 0|87|96| 43|16|5 9|37|39| <u>31</u>|68|5 9|71|50

45|74|0 4|<u>18</u>|<u>07</u>| 65|56|<u>1 3</u>|<u>33</u>|<u>02</u>| 07|05|1 9|36|<u>23</u>| 18|13|2 0|95|47

<u>27</u>|

giving the treatment 2 group as Tullock, Vaughn, Liang, Durr, Howard, Wilder, Bennett, Ogle, and Solomon. Continuing in line 134, the treatment 3 group is

|81|6 7|84|16| 18|32|9 2|13|37| <u>35</u>|<u>21</u>|3 3|77|41| 04|31|<u>2 6</u>|85|<u>08</u>

66|92|5 5|56|58| 39|<u>10</u>|0 7|84|58| <u>11</u>|20|6 1|98|76| 87|<u>15</u>|1 3|<u>12</u>|60

08|42|<u>1 4</u>|

Zhang, Montoya, Rosen, Edwards, Fleming, George, Imrani, Han, and Hruska assigned to treatment 3. The nine remaining subjects, Ashiro, Bikalis, Duncan, Farouk, O'Brian, Padilla, Trujillo, Valasco, and Willis are assigned to treatment 4. Using Table B can become tedious with a large number of subjects and it is best to leave such calculations to a computer.

Exercise 3.29

We have 15 subjects available. The factor is "hand" and there are two treatments. Treatment 1 is squeezing with the right hand and treatment 2 is squeezing with the left hand. The response is the force exerted as indicated by the reading on the scale.

To do the experiment, we use a matched pairs design. Each subject is a block and each subject uses each hand (the matched pairs are the two hands of a particular subject). We should randomly decide which hand to use first, perhaps by flipping a coin. We measure the response for each hand and then compare the forces for the left and right hands over all subjects to see if there is a systematic difference between the two hands.

If we use a coin to do the randomization, we might decide that if we get heads the right-hand goes first. Tails means the left-hand goes first. For 15 flips of a coin we got

HTTHTTHTHHHTTTT

This tells us that subject 1 uses the right-hand first, subject 2 the left hand first, subject 3 the left-hand first, etc. Notice with this scheme the number of times the right-hand goes first is 6 and the number of times the left-hand goes first is 9.

Exercise 3.32

a) The subjects and their excess weights, rearranged in increasing order of excess weight, are listed below. The columns are the five blocks. We have labeled the subjects in each block from 1 to 4.

Block 1	Block 2	Block 3	Block 4	Block 5
1 Williams 22	1 Santiago 27	1 Brunk 30	1 Jackson 33	1 Birnbaum 35
2 Festinger 24	2 Mann 28	2 Obrach 30	2 Stall 33	2 Tran 35
3 Hernandez 25	3 Smith 29	3 Rodriguez 30	3 Brown 34	3 Nevesky 39
4 Moses 25	4 Kendall 30	4 Loren 32	4 Dixon 34	4 Wilansky 42

b) We used lines 130 and 131 in Table B, which are given below.

69051 64817 87174 09517 84534 06489 87201 97245

05007 16632 81194 14873 04197 85576 45195 96565

For block 1, we read these from left to right, one digit at a time. The first label we encounter is assigned to regimen A, the next to regimen B, the next to regimen C, and the remaining label is then automatically assigned to regimen D. We underline those that are one of our labels, skipping repeats. The vertical lines indicates when we have completed a block. We summarize our results on the next page.

Regimen A = Williams, Kendall, Obrach, Brown, Birnbaum

Regimen B =Moses, Mann, Loren, Stall, Wilansky

Regimen C = Hernandez, Santiago, Brunk, Jackson, Nevesky

Regimen D = Festinger, Smith, Rodriguez, Dixon, Tran

Exercise 3.35

a) This is false. Randomness does not mean each digit appears the exact same number of times in each row. For example, look at line 150 in Table B. There are only two 0s in this row.

b) True. Randomness means each digit has an equal chance (1/10) of being a 0, each pair an equal chance (1/100) of being 00, each triple an equal chance (1/1000) of being 000, etc. This point is made in the section "how to randomize" in the text a few paragraphs below the box containing the definition of a table of random digits.

c) False. Following the logic in (b), any set of four digits has an equal chance (1/10000) of being 0000. Thus the digits 0000 can appear, but the chance of any four digits being this sequence is quite small.

SECTION 3.3

OVERVIEW

The **population** is the entire group of individuals or objects about which we want information. The information collected is contained in a **sample** which is the part of the population we actually get to observe. How the sample is chosen, that is, the **design**, has a large impact on the usefulness of the data. A useful sample will be representative of the population and will help answer our questions. "Good" methods of collecting a sample include the following:

> **probability samples**
> **simple random samples,** also called **SRS**
> **stratified random samples**
> **multistage samples**

All these sampling methods involve some aspect of randomness through the use of a formal chance mechanism. Random selection is just one precaution that a person can take to reduce **bias,** the systematic favoring of a certain outcome. Samples we select using our own judgment, because they are convenient, or "without forethought" (mistaking this for randomness) are usually biased in some way. This is why we use computers or a tool like a **table of random digits** to help us select a sample.

A **voluntary response sample** includes people who choose to be in the sample by responding to a general appeal. They tend to be biased, as the sample is overrepresented by individuals with strong opinions, which are often negative.

Other kinds of bias to be on the lookout for include:

nonresponse bias which occurs when individuals who are selected do not participate or cannot be contacted,

undercoverage which occurs when some group in the population is given either nochance or a much smaller chance than other groups to be in the sample, and

response bias which occurs when individuals do participate but are not responding truthfully or accurately due to the way the question is worded, the presence of an observer, fear of a negative reaction from the interviewer, or any other such source.

These types of bias can occur even in a randomly chosen sample and we need to try to reduce their impact as much as possible.

GUIDED SOLUTIONS

Exercise 3.39

KEY CONCEPTS - populations and sources of bias

What variable was measured and what was the sample? Now, try and identify the population as exactly as possible. Where the information is not complete, you may need to make assumptions to try to describe the population in a reasonable way. Make sure not to confuse the population of interest with the population actually sampled. When they don't coincide there is always a strong potential for bias. What are some possible sources of bias in this example?

Exercise 3.41

KEY CONCEPTS - selecting a SRS with a table of random numbers

The table of random numbers can be used to select a SRS of numbers - in order to use it to sample from the students in the statistics course, the individuals in the course need to be assigned numbers. So that everyone does the problem the "same" way, we have first numbered the students according to alphabetical order in the list.

01- Agarwal	08 - Dewald	15 - Huang	22 - Puri
02 - Alfonseca	09 - Fleming	16 - Kim	23 - Richards
03 - Baxter	10 - Fonseca	17 - Lujan	24 - Rodriguez
04 - Bowman	11 - Gates	18 - Mourning	25 - Santiago
05 - Brown	12 - Goel	19 - Nunez	26 - Shen
06 - Cortez	13 - Gomez	20 - Peters	27 - Vega
07 - Cross	14 - Hernandez	21 - Pliego	28 - Watanabe

If you go to line 139 in the table and start selecting two digit numbers, then you should get the same answer as given in the complete solution.

Exercise 3.46

KEY CONCEPTS - systematic sampling

a) This is like the example except there are now 200 addresses instead of 100, and the sample size is now 5 instead of 4. With these two changes, you need to think about how many different systematic samples there are. Two different systematic samples are:

systematic sample 1 =   01, 41, 81, 121, 161
systematic sample 2 =   02, 42, 82, 122, 162
....

How many systematic samples are there altogether? Choosing one of these systematic samples at random is equivalent to choosing the first address in the sample. The remaining four addresses follow automatically by adding 40. Carry this out using line 120 in the table.

b) Why are all addresses equally likely to be selected? First, how many systematic samples contain each address? The chance of selecting an address is the same as the chance of selecting the systematic sample that contains it. With this in mind, what is the chance of any address being chosen? By the definition of a SRS, all samples of 5 addresses are equally likely to be selected. In a systematic sample, are all samples of 5 addresses even possible?

Exercise 3.51

KEY CONCEPTS - sampling frame, undercoverage

a) Which households wouldn't be in the sampling frame? Make some educated guesses as to how these households might differ from those in the sampling frame (other than the fact that they don't have a phone number in the directory).

b) Random digit dialing makes the sampling frame larger - which households are added to it?

Exercise 3.55

KEY CONCEPTS - wording of questions

Questions can be worded in such a way that makes it seem as though any reasonable person should agree (disagree) with the statement. Which questions are slanted towards a desired response? Are all the questions clear?

COMPLETE SOLUTIONS

Exercise 3.39

The variable being measured is approval of the president's overall job performance which is recorded as approve or don't approve. The sample is the 1210 adults that were actually interviewed. The population of interest is probably all adult citizens of the U.S. or possibly just registered voters.

There are several possible sources of bias in the study. The states of Alaska and Hawaii were omitted and there is no reason to believe that the adult residents of these states were not intended to be part of the population (they may not have been included in the sample due to the higher cost of calling residents of these states). Any systematic differences in the opinions of the adults in Alaska and Hawaii and the remaining states will bias the results. Also, only residents with phones could be contacted and if the phone numbers were selected from phone books then residents with unlisted numbers could not be in the sample. This is another possible source of bias, which is just any systematic error in the way the sample represents the population. Finally, there may be bias due to nonresponse, as all adults contacted by phone may not have been willing to give their opinion.

Exercise 3.41

To choose a SRS of 6 students to be interviewed, first label the members of the population by associating a 2 digit number with each.

01- Agarwal	08 - Dewald	15 - Huang	22 - Puri
02 - Alfonseca	09 - Fleming	16 - Kim	23 - Richards
03 - Baxter	10 - Fonseca	17 - Lujan	24 - Rodriguez
04 - Bowman	11 - Gates	18 - Mourning	25 - Santiago
05 - Brown	12 - Goel	19 - Nunez	26 - Shen
06 - Cortez	13 - Gomez	20 - Peters	27 - Vega
07 - Cross	14 - Hernandez	21 - Pliego	28 - Watanabe

Now enter Table B and read two-digit groups until 6 students are chosen. Starting at line 139

55588 99404 70708 41098 43563 56934 48394 51719
12975 13258 13048

The selected sample is 04 - Bowman, 10 - Fonseca, 17 - Lujan, 19 - Nunez, 12 - Goel, and 13 - Gomez.

Exercise 3.46

a) We want to select 5 addresses out of 200, so we think of the 200 addresses as forty lists, each containing 5 addresses. We choose one address from the first 40, and then every 40th address after that. The first step is to go to Table B, line 120 and choose the first two digit random number you encounter that is one of the numbers 01, ..., 40.

35476

The selected number is 35, so the sample includes addresses numbered 35, 75, 115, 155, and 195.

b) Each individual is in exactly one systematic sample, and the systematic samples are equally likely to be chosen. In our previous example, there were 40 systematic samples, each containing 5 addresses. The chance of selecting any address is the chance of picking the systematic sample that contains it, which is 1 in 40.

A simple random sample of size n would allow every set of n individuals an equal chance of being selected. Thus, in this exercise, when using a SRS the sample consisting of the addresses numbered 1, 2, 3, 4, and 5 would have the same probability of being selected as any other set of 5 addresses. For a systematically selected sample, all samples of size n do not have the same probability of being selected. In our exercise the sample consisting of the addresses numbered 1, 2, 3, 4, and 5 would have zero chance of being selected since the numbers of the addresses do not all differ by 40. The sample we selected in (a), 35, 75, 115, 155, and 195 had a 1 in 40 chance of being selected, so all samples of five addresses are not equally likely.

Exercise 3.51

a) Households omitted from the frame are those which do not have a telephone number listed in the telephone directory. The types of people who might be underrepresented are poorer (including homeless) people who cannot afford to have a phone, and the group of people who have unlisted numbers. It is harder to characterize this second group. As a group they would tend to have more money as you need to pay to have your phone number unlisted or it might include more single women who do not want their phone numbers available and possibly people whose jobs put them in contact with large groups of people who might harass them if their phone number was easily accessible.

b) People with unlisted numbers will be included in the sampling frame. The sampling frame would now include any household with a phone. One interesting point is that all households will not have the same probability of getting in the sample, as some households have multiple phone lines and will be more likely to get in the sample. So, strictly speaking, random digit dialing will not actually provide a SRS of households with phones. Just a SRS of phone numbers!

Exercise 3.55

a) The beginning of the question suggests that cell phone use is associated with brain cancer. This initial suggestion and the wording "the danger of using cell phones" would lead most reasonable people to be in favor of including a warning label. The question is slanted in favor of this response.

b) The question is clear but is slanted in favor of national health insurance. The reason for agreeing with a question should not be contained within the question.

c) The question is slanted as it contains reasons why you should support recycling. As a question, the wording is a little technical for the general population and a simpler version such as "Do you favor economic incentives to promote recycling?" would be better.

SECTION 3.4

OVERVIEW

Statistical inference is the technique which allows us to use the information in a sample to draw conclusions about the population. To understand the idea of statistical inference, it is important to understand the distinction between **parameters** and **statistics**. A **statistic** is a number we calculate based on a sample from the population - its value can be computed once we have taken the sample, but its value varies from sample to sample. A statistic is generally used to estimate a population **parameter** which is a fixed but unknown number that describes the population.

The variation in a statistic from sample to sample is called **sampling variability**. It can be described through the **sampling distribution** of the statistic which is the distribution of values taken by the statistic in all possible samples of the same size from the population. The sampling distribution can be described in the same way as the distributions we encountered in Chapter 1. Three important features are:

- a measure of center
- a measure of spread
- a description of the shape of the distribution

The properties and usefulness of a statistic can be determined by considering its sampling distribution. If the sampling distribution of a statistic is centered (has its mean) at the value of the population parameter, then the statistic is **unbiased** for this parameter. This means that the statistic tends to neither overestimate nor underestimate the parameter.

Another important feature of the sampling distribution is its spread. If the statistic is unbiased and the sampling distribution has little spread or variability, then the statistic will tend to be close to the parameter it is estimating for most samples. The variability of a statistic is related to both the sampling design and the sample size n. Larger sample sizes give smaller spread (better estimates) for any sampling design. An important feature of the spread is that as long as the population is much larger than the sample (at least 100 times), the spread of the sampling distribution will depend primarily on the sample size, not the population size.

If the parameter **p** is the proportion of the population with a particular characteristic, then the statistic $\hat{p}$, the proportion in the sample with this characteristic, is an unbiased estimator. Provided the samples are selected at random, **probability** theory can be used to tell us about the distribution of a statistic.

GUIDED SOLUTIONS

Exercise 3.59

KEY CONCEPTS - statistics and parameters

In deciding whether a number represents a parameter or a statistic, you need to think about whether it is a numerical characteristic of the population of interest or whether it is a numerical characteristic of the particular sample that was selected. Statistics vary from sample to sample; parameters are fixed numerical characteristics of the population.

Exercise 3.65

KEY CONCEPTS - variability of the sample proportion.

a) "As long as the population is much larger than the sample (say, at least 100 times as large), the spread of the sampling distribution for a sample of fixed size n is approximately the same for any population size." You need to think about how this rule applies to this example.

b) Is the rule given in part (a) applicable here? Read it carefully.

Exercise 3.69

KEY CONCEPTS - sampling distributions

a) The table of random numbers contains the 10 digits, 0, 1, 2, ..., 9, which are "equally likely" to occur in any position selected at random from the table. If we want an egg mass to be present 20% of the time, then two digits correspond to the presence of an egg mass and the remaining eight digits correspond to the absence of an egg mass. Does it matter which two digits correspond to the presence of an egg mass?

While any two digits could be used, so that everyone does the same thing, let the occurence of the digits 0 or 1 correspond to the presence of an egg mass, and the remaining digits correspond to the absence. Also, let's start on line 128 of the table.

1̲5689 1̲4227

These 10 random digits correspond to our 10 sample areas. There are two sample areas with egg masses (correspond to a digit of 0 or 1), so that $\hat{p} = .2$ for this sample.

b) For this part of the problem, everyone will be taking their 20 samples from different parts of the random number table. Some of you may know how to get random samples from your computer software. Work with your own samples here. Your answers will not agree exactly with that given in the complete solution - the general pattern should be similar. If everyone took 2000 samples instead of 20, would the sampling distributions from person to person show more or less agreement?

COMPLETE SOLUTIONS

Exercise 3.59

2.503 cm. is a property (the mean) of the carload (population) of ball bearings and is the value of a parameter. **2.515** cm. is a property of the sample of 100 bearings inspected. It is the value of a statistic.

Exercise 3.65

a) The population is at least 100 times the sample size $n = 2000$ for each of the states. So the variablility in the sample proportion based on $n = 2000$ will be approximately the same for the population size of any state.

b) The problem switches here. The rule applies to a given sample size - the variability of the sample proportion based on a fixed number of observations is approximately the same for any population size. Now the sample size will vary from state to state. For Wyoming, 1/10 of 1% of the population is a sample size of about n = 494 and 1/10 of 1% of the population of California is a sample size of about n = 34000. Since larger sample sizes give smaller spread, there will be differences in the variability of the sample proportion from state to state. California's sample proportion will be much less variable than the sample proportion from Wyoming.

Exercise 3.69

a) Done in the guided solution

b) These are the values of $\hat{p}$ in the 20 samples we obtained using the computer to generate random digits, followed by the stem and leaf plot.

sample	$\hat{p}$
1	0.1
2	0.1
3	0.0
4	0.1
5	0.0
6	0.4
7	0.0
8	0.3
9	0.2
10	0.0
11	0.2
12	0.0
13	0.4
14	0.2
15	0.4
16	0.2
17	0.2
18	0.1
19	0.3
20	0.2

```
0.0 |00000
0.1 |0000
0.2 |000000
0.3 |00
0.4 |000
```

The mean of the distribution is 0.17. The shape looks fairly symmetric with a center near 0.2. Your stem and leaf plot may look quite different from this - with only 20 samples the distributions may vary quite a bit from person to person. If everyone took 2000 samples, which would require the sampling be done using a computer, then the shapes of the distributions would be quite similar from person to person.

CHAPTER 4

PROBABILITY: THE STUDY OF RANDOMNESS

SECTION 4.1

OVERVIEW

A process or phenomenon is called **random** if its outcome is uncertain. Although individual outcomes are uncertain, when the process is repeated a large number of times the underlying distribution for the possible outcomes begins to emerge. For any outcome, its **probability** is the proportion of times, or the relative frequency, with which the outcome would occur in a long series of repetitions of the process. It is important that these repetitions or trials be **independent** for this property to hold.

You can study random behavior by carrying out physical experiments such as coin tossing or rolling of a die, or you can simulate a random phenomenon on the computer. Using the computer is particularly helpful when we want to consider a large number of trials.

GUIDED SOLUTIONS

Exercise 4.3

KEY CONCEPTS - random phenomena

This is a good opportunity to observe chance variation. When tossing a thumbtack 100 times, although each student will have the point face up on different tosses, the proportion of times in which the point faces up will not differ that much from student to student. In Chapter 5 you will learn how to calculate how much this proportion will vary from student to student. However, for now simply record the number of times the point faces up in the space below and use this to approximate the probability of the thumbtack landing point up. Be sure to use a hard surface as the probability of landing point up will be different if you toss the thumbtack on a carpet.

Number of times landing point up =

Approximate probability =

Exercise 4.7

KEY CONCEPTS - simulating a random phenomenon

a) You will need to use your computer software to simulate the 100 trials or the Applet. After simulating the 100 trials calculate the proportion of "hits."

proportion of hits =

For most students, their proportion of hits will be within 0.05 or 0.10 of the true probability of .5.

b) You need to go through your sequence to determine the longest string of hits or misses.

Longest run of shots hit =   Longest run of shots missed =

COMPLETE SOLUTIONS

Exercise 4.3

Our 100 tosses yielded the following results. Your answers will not agree exactly, but may disagree due to the brand of thumbtack you are using as well as chance variation   The length of the point and the size of the head of the thumbtack will vary from brand to brand, and these factors will affect the probability of landing point up. These difficulties do not arise when tossing pennies, as different pennies are much more similar to each other.

Number of times landing point up = 55

Approximate probability = 55/100 = 0.55 = 55%

Exercise   4.7

a) Our sequence of hits (H) and misses (M) is given below.

```
H   H   M   H   H   H   M   M   H   H   H   M   H   M   H
M   H   M   M   H   M   M   H   H   H   M   M   H   H   H
M   M   M   M   H   M   M   H   H   H   H   M   H   H   H
M   M   M   M   M   H   M   H   H   M   H   M   H   M   M
H   H   H   H   M   H   M   M   M   M   H   H   M   H   H
M   M   H   H   H   M   M   H   H   M   M   H   M   H   M
M   H   M   H   H   M   H   H   H   H
```

proportion of hits = .54

b) You need to go through your sequence to determine the longest string of hits or misses. In our example,

Longest run of shots hit = 4 (this occurred more than once)
Longest run of shots missed = 5

SECTION 4.2

OVERVIEW

The description of a random phenomenon begins with the **sample space** which is the list of all possible outcomes. A set of outcomes is called an **event.** Once we have determined the sample space, a **probability model** tells us how to assign probabilities to the various events that can occur. There are four basic rules that probabilities must satisfy.

- Any probability is a number between 0 and 1.
- All possible outcomes together must have probability 1.
- The probability that an event does not occur is 1 minus the probability that the event occurs.
- If two events have no outcomes in common, the probability that one or the other occurs is the sum of their individual probabilities.

In a sample space with a finite number of outcomes, probabilities are assigned to the individual outcomes and the probability of any event is the sum of the probabilities of the outcomes that it contains. In some special cases, the outcomes are all **equally likely** and the probability of any event A is just computed as

$P(A)$ = (number of outcomes in A)/(number of outcomes in S).

Events are **disjoint** if they have no outcomes in common. In this special case the probability that one or the other event occurs is the sum of their individual probabilities. This is the addition rule for disjoint events, namely

$$P(A \text{ or } B) = P(A) + P(B).$$

Events are **independent** if knowledge that one event has occurred does not alter the probability that the second event occurs. The mathematical definition of independence leads to the **multiplication rule** for independent events. If A and B are independent, then

$$P(A \text{ and } B) = P(A)P(B).$$

In any particular problem we can use this definition to check if two events are independent by seeing if the probabilities multiply according to the definition. However, most of the time, independence is assumed as part of the probability model. The four basic rules, plus the multiplication rule, allow us to compute the probabilities of events in many random phenomena.

Many students confuse independent and disjoint events once they have seen both definitions. Remember, disjoint events have no outcomes in common and when two events are disjoint, you can compute $P(A \text{ or } B) = P(A) + P(B)$ in this special case. The probability being computed is that one or the other event occurs. Disjoint events cannot be independent since once we know that A has occurred, then the probability of B occurring becomes 0 (B cannot have occurred as well - this is the meaning of disjoint). The multiplication rule can be used to compute the probability that two events occur simultaneously, $P(A \text{ and } B) = P(A)P(B)$, in the special case of independence.

GUIDED SOLUTIONS

Exercise 4.11

KEY CONCEPTS - sample space

One of the main difficulties encountered when describing the sample space is finding some notation to express your ideas formally. Following the text, our general format is $S = \{\quad\}$, where a description of the outcomes in the sample space is included within the braces.
a) You want to express that any number between 0 and 24 is a possible outcome. So you would write $S = \{$all numbers between 0 and 24$\}$.

b) You may not want to put an upper bound on the amount, so you should allow for any number greater than 0, remembering that certain numbers are not possible values for the amount of change a student is carrying.

$S =$

c) $S =$

d) $S =$

Exercise 4.13

KEY CONCEPTS - applying the probability rules

a) Since these are the only blood types, what has to be true about the sum of the probabilities for the different types? Use this to find $P(AB)$.

b) What's true about the events O and B blood type? Which probability rule do we follow? (Don't be confused by the wording in the problem which says "people with blood types O <u>and</u> B." In the context of this problem and the language of probability we are using, it really means O or B. There are no people with blood types O and B).

Exercise 4.21

KEY CONCEPTS - independent events

Although independence is often assumed in setting up a probability model, in other cases we must use the formal definition to determine if two events are independent. In order to determine if the events A = Hispanic and B = White are independent, we must see if they satisfy the multiplication rule. This requires that we evaluate $P(A)$, $P(B)$ and $P(A$ and $B)$. To evaluate $P(A)$ we need to add up the proportions of the population corresponding to each race that are also Hispanic.

$$P(A) = 0.000 + 0.003 + 0.060 + 0.062 = 0.125.$$

Now evaluate $P(B)$ and $P(A$ and $B)$ on your own, and determine if the multiplication rule is satisfied for these events.

$P(B) =$
$P(A$ and $B) =$
$P(A)P(B) =$

Exercise 4.29

KEY CONCEPTS - sample spaces for simple random sampling, probabilities of events

a) It's easy to make the list. S = {(Abby, Mei-Ling), (Abby, Julie), etc.}. There is no need to include both (Abby, Mei-Ling) and (Mei-Ling, Abby) in your list, since both refer to the same two individuals.
b) How many outcomes are there in the sample space in (a)? If they are equally likely, what is the probability of each?

c) How many outcomes in *S* include Mei-Ling? When the outcomes are equally likely, the probability of the event is just the

(number of outcomes in the event) / (number of outcomes in *S*).

d) How many outcomes in *S* include neither of the two men?

Exercise 4.31

KEY CONCEPTS: independence, multiplication rule

The probability of winning the major battle is 0.6. What is the probability of winning all three small battles? How would you decide which strategy is best? Write the event "winning all three small battles" in terms of winning each of the small battles and use the independence of victories or defeats in the small battles.

P(winning all three small battles) =

Which strategy do you prefer and why?

Exercise 4.35

KEY CONCEPTS - multiplication rule for independent events

a) The three years are independent. If *U* indicates a year for the price being up and *D* indicates a year for the price being down, you need to compute $P(UUU)$.

b) Since the events are independent, what happens in the first two years does not affect the probability of going up or down in the third year. What's the probability of the price going down in any given year?

c) This problem must be set up carefully and done in steps.

Step 1 - Write the event of interest in terms of simpler outcomes. How would you write $P(UU$ or $DD)$ in terms of $P(UU)$ and $P(DD)$?

P(moves in the same direction in the next two years) = $P(UU$ or $DD)$.

Step 2 - Evaluate $P(UU)$ and $P(DD)$ and substitute your answer in the expression from Step 1.

COMPLETE SOLUTIONS

Exercise 4.11

a) $S = \{$all numbers between 0 and 24$\}$.
b) $S = \{0, 0.01, 0.02, 0.03, ...\}$, where 0.01 corresponds to 1 cent, etc.
c) $S = \{$A, B, C, D, F$\}$. You may need to include +'s and -'s depending on the grading scheme for your school.
d) $S = \{$no, yes$\}$.

Exercise 4.13

a) The probabilities for the different blood types must add to 1. The sum of the probabilities for blood types O, A, and B is $0.45 + 0.40 + 0.11 = 0.96$. Subtracting this from 1 tells us that the probability of the remaining type AB must be 0.04.

b) Maria can receive transfusions from people with blood types O or B. Since a person cannot have both of these blood types, they are disjoint. The calculation follows probability rule 4 which says $P(\text{O or B}) = P(\text{O}) + P(\text{B}) = 0.45 + 0.11 = 0.56$.

Exercise 4.21

$$P(B) = 0.060 + 0.691 = 0.751.$$

The probability of being both white and Hispanic corresponds to the single entry in the column labeled "Hispanic" and the row labeled "white." Reading the entry in the table gives $P(A \text{ and } B) = 0.060$. Since $P(A)P(B) = 0.125 \times 0.751 = 0.094$, the multiplication rule is not satisfied and the events are not independent.

Exercise 4.29

a) $S = \{$(Abby, Mei-Ling), (Abby, Julie), (Abby, Sam), (Abby, Roberto), (Mei-Ling, Julie), (Mei-Ling, Sam), (Mei-Ling, Roberto), (Julie, Sam), (Julie, Roberto), (Sam, Roberto)$\}$.

b) There are 10 possible outcomes. Since they are equally likely, each has probability 0.10.

c) Mei-Ling is in 4 of the outcomes so her chance of attending the conference in Paris is 4 / 10 = 0.4. (Note that each person has the same probability of going. Remember from Chapter 2 this is a property of a SRS).

d) The chosen group must contain two women, (Abby, Mei-Ling), (Abby, Julie), or (Mei-Ling, Julie). There are three possibilities, so the desired probability is 3 / 10 = 0.3.

Exercise 4.31

Denote the event that the general wins the first small battle by W_1, the event that the general wins the second small battle by using W_2, and the event that the general wins the third small battle by using W_3. Then

$$P(\text{winning all three small battles}) = P(W_1 \text{ and } W_2 \text{ and } W_3) = P(W_1)P(W_2)P(W_3)$$

since victories or defeats in the small battles are independent. We know that the probability of winning each small battle is 0.8, so

$$P(\text{winning all three small battles}) = P(W_1)P(W_2)P(W_3) = 0.8 \times 0.8 \times 0.8 = 0.512.$$

Because the general is more likely to win the major battle than to win all three small battles, his strategy should be to fight one major battle.

Exercise 4.35
a) $P(UUU) = (0.65)^3 = 0.2746$

b) The probability of the price being down in any given year is $1 - 0.65 = 0.35$. Since the years are independent, the probability of the price being down in the third year is 0.35, regardless of what has happened in the first two years.

c) $P(\text{moves in the same direction in the next two years}) = P(UU \text{ or } DD) = P(UU) + P(DD)$, since the events UU and DD are disjoint. Using the independence of two successive years, $P(UU) = (0.65)^2 = 0.4225$, and $P(DD) = (0.35)^2 = 0.1225$. Putting this together,

$P(\text{moves in the same direction in the next two years}) = 0.4225 + 0.1225 = 0.5450.$

SECTION 4.3

OVERVIEW

A **random variable** is a variable whose value is a numerical outcome of a random phenomenon. The restriction to numerical outcomes makes the description of the probability model simpler and allows us to begin to look at some further properties of probability models in a unified way. If we toss a coin three times and record the sequence of heads and tails, then an example of an outcome would be HTH, which would not correspond directly to a random variable. On the other hand, if we were only keeping track of the number of heads on the three tosses, then the outcome of the experiment would be 0, 1, 2, or 3 and would correspond to the value of the random variable X = number of heads.

The two types of random variables we will encounter are **discrete** and **continuous** random variables. The **probability distribution** of a random variable tells us about the possible values of X and how to assign probabilities to these values. A discrete random variable has a finite number of values, and the probability distribution is a list of the possible values of X and the probabilities assigned to these values. The probability distribution can be given in a table or using a **probability histogram**. For any event described in terms of X, the probability of the event is just the sum of the probabilities of the values of X included in the event.

A continuous random variable takes all values in some interval of numbers. Probabilities of events are determined using a **density curve**. The probability of any event is the area under the curve corresponding to the values that make up the event. For density curves that involve regular shapes such as rectangles or triangles, we can compute probabilities of events using simple geometrical arguments. The **normal distribution** is another example of a continuous probability distribution, and probabilities of events for normal random variables are computed by standardizing and referring to Table A as was done in Section 1.3.

GUIDED SOLUTIONS

Exercise 4.45

KEY CONCEPTS - discrete random variables, computing probabilities

a) Write the event in terms of a probability about the random variable X. While you can figure out the answer without doing this, it's good practice to start using the notation for random variables. To find the probability that "the unit has 5 or more rooms," add the appropriate probabilities given in Exercise 4.43 that provides the distribution of X. (Be sure to use the row for owned units).

b) Express the event in words and compute its probability using the distribution of X in Exercise 4.43. How is this event different than the event in part (a).

c) You should have different answers in part (a) and part (b). What fact about discrete random variables does this illustrate? If X had a continuous distribution, would the events in (a) and (b) have the same probability?

Exercise 4.50

KEY CONCEPTS - finding the probability distribution of a random variable

a) The probability of a randomly selected student opposing the funding of interest groups is 0.4 and the probability of favoring it is 0.6. The opinions of different students sampled are independent of each other. So you can use the multiplication rule to find

$P(A$ supports, B supports, and C opposes) =

b) It is easiest to do this by making a table to keep track of the calculations. The first entry is given below. There should be eight lines to the table when you're done. If you've done the calculations correctly, what should be true about the eight probabilities? Don't worry about the column labeled value of X for now. It will not be needed until part (c).

A	B	C	Probability	Value of X
support	support	support	$(0.6)^3 = 0.216$	

c) For each committee listed in the table in (b), find the associated value of X. For the first row, 0 people oppose the funding of interest groups so the value of X is 0 for a committee with these views. The values of X which can occur are 0, 1, 2, and 3. To find the probability that X takes any of these values, just add up the probabilities of the committees with that value of X. Fill in the table below with your values and make sure that the probabilities sum to 1.

Value of X	Probability

d) If a majority oppose funding, how many people on the committee would have to oppose funding? What does this say about X? Now use the table you constructed in (c) to evaluate this probability.

Exercise 4.53

KEY CONCEPTS - continuous random variables, computing probabilities

a) The graph of the density curve is given below. The property you are given is that the density has a constant height between 0 and 2. Since the area under the density curve must be equal to 1, what is the height? Recall that the area of a rectangle is the length × height.

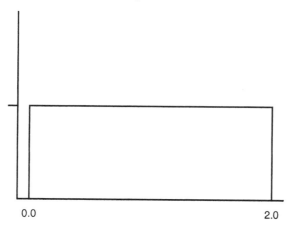

b) As with finding areas under normal curves, it helps to draw a sketch of the density which includes the area corresponding the probability that you need to evaluate. In this part you need to find $P(Y \le 1)$, when Y is a random number between 0 and 2. The density curve with the area corresponding to this probability is given below. Since it is a rectangular region, the area corresponds to the length × height = 0.5 × 1 = 0.5, which is the $P(Y \le 1)$. Remember for continuous densities, $P(Y \le 1) = P(Y < 1)$.

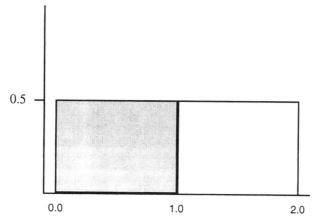

c) Sketch the density and the area you need below.

d) Sketch the density and the area you need below.

Exercise 4.55

KEY CONCEPTS - probabilities for a sample proportion

a) Finding probabilities associated with a sample proportion when we know the mean and standard deviation of the sampling distribution, requires first standardizing to a z-score so that we can refer to the table for the standard normal distribution. In this example the mean is $\mu = 0.30$, (the value of p), and the standard deviation is given as $\sigma = 0.023$. If you are still uncomfortable doing this type of problem, it is best to continue to draw a picture of a normal curve and the required area as we did in Section 1.3. Otherwise you can just follow the method of Example 4.18 in this chapter. At least half the sample corresponds to the sample proportion being greater than or equal to 0.5.

$P(\hat{p} \geq 0.5) =$

b) $P(\hat{p} < 0.25) =$

c) $P(0.25 < \hat{p} < 0.35) =$

COMPLETE SOLUTIONS

Exercise 4.45

a) $P(X \geq 5) = 0.210 + 0.224 + 0.197 + 0.149 + 0.053 + 0.035 = 0.868$

b) The event $\{X > 5\}$ indicates that "the unit has more than 5 rooms." The probability is computed as

$P(X > 5) = 0.224 + 0.197 + 0.149 + 0.053 + 0.035 = 0.658$

c) The answers are different because for discrete distributions the probability of an individual outcome is not necessarily zero. In this case, the probability of the individual outcome 5 is not zero, so that $P(X > 5)$ does not have the same value as $P(X \geq 5)$. (Note: For this distribution of X, $P(X > 4.5) = P(X \geq 4.5)$ since the outcome 4.5 has zero probability).

Exercise 4.50

a) P(A supports, B supports and C opposes)
$\quad\quad = P$(A supports) P(B supports) P(C opposes) $= (0.6)(0.6)(0.4) = 0.144$

b)

A	B	C	Probability	Value of X
support	support	support	$(0.6)^3 = 0.216$	0
support	support	oppose	$(0.6)^2(0.4) = 0.144$	1
support	oppose	support	$(0.6)^2(0.4) = 0.144$	1
oppose	support	support	$(0.6)^2(0.4) = 0.144$	1
support	oppose	oppose	$(0.6)(0.4)^2 = 0.096$	2
oppose	support	oppose	$(0.6)(0.4)^2 = 0.096$	2
oppose	oppose	support	$(0.6)(0.4)^2 = 0.096$	2
oppose	oppose	oppose	$(0.4)^3 \quad = 0.064$	3
			Total $\quad = 1.00$	

c) The value of X is given in the table in (b). The possible values of X are 0, 1, 2, and 3. To find the probability that X takes any value, just add up the probabilities of the committees with that value of X. For example, $P(X = 2) = 3(0.096) = 0.288$.

Value of X	Probability
0	0.216
1	0.432
2	0.288
3	0.064

d) A majority oppose say that either 2 or 3 members of the board oppose. If 2 oppose then $X = 2$ and if 3 oppose then $X = 3$. So the event is $X \geq 2$ and the required probability is $P(X \geq 2) = 0.288 + 0.064 = 0.352$.

Exercise 4.53

a) The area is the length $\times$ height. Since the area is 1 and we know the length is 2, you must solve the equation $2 \times$ height $= 1$. Thus the height must be equal to 1/2 or 0.5.

b) See the Guided Solutions.

c) The shaded area is $P(0.5 < Y < 1.3) = (0.8)(0.5) = 0.40$.

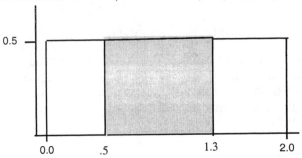

d) The shaded area is $P(Y \geq 0.8) = (1.2)(0.5) = 0.60$.

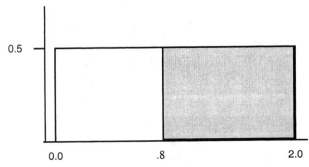

Exercise 4.55

a) $\qquad P(\hat{p} \geq 0.50) = P\left(\dfrac{\hat{p} - 0.30}{0.023} \geq \dfrac{0.50 - 0.30}{0.023}\right) = P(Z \geq 8.70) \approx 0$

b) $\qquad P(\hat{p} < 0.25) = P\left(\dfrac{\hat{p} - 0.30}{0.023} \leq \dfrac{0.25 - 0.30}{0.023}\right) = P(Z \leq -2.17) = 0.015$

c) $\qquad P(0.25 < \hat{p} < 0.35) = P\left(\dfrac{0.25 - 0.30}{0.023} \leq \dfrac{\hat{p} - 0.30}{0.023} \leq \dfrac{0.35 - 0.30}{0.023}\right)$

$$= P(-2.17 \leq Z \leq 2.17) = 0.985 - 0.015 = 0.970.$$

SECTION 4.4

OVERVIEW

In Chapter 1 we introduced the concept of the distribution of a set of numbers or data. The distribution describes the different values in the set and the frequency or relative frequency with which those values occur. The mean of the numbers is a measure of the center of the distribution and the standard deviation is a measure of the variability or spread. These concepts are also used to describe features of a random variable X. The probability distribution of a random variable indicates the possible values of the random variable and the probability (relative frequency in repeated observations) with which they occur.

The **mean** μ_X of a random variable X describes the center or balance point of the probability distribution or density curve of X. If X is a discrete random variable having possible values x_1, x_2, ...,x_k with corresponding probabilities $p_1, p_2, ...,p_k$, the mean μ_X is the average of the possible values weighted by the corresponding probabilities, i.e.

$$\mu_X = x_1 p_1 + x_2 p_2 + ... + x_k p_k$$

The mean of a continuous random variable is computed from the density curve but computations require more advanced mathematics. The law of large numbers relates the mean of a set of data to the mean of a random variable and says that the average of the values of X observed in many trials approaches μ_X.

The **variance** σ_X^2 of a random variable X is the average squared deviation of the values of X from their mean. For a discrete random variable

$$\sigma_X^2 = (x_1 - \mu_X)^2 p_1 + (x_2 - \mu_X)^2 p_2 + ... + (x_k - \mu_X)^2 p_k.$$

The **standard deviation** σ_X is the positive square root of the variance. The standard deviation measures the variability of the distribution of the random variable X about its mean. The variance of a continuous random variable, like the mean, is computed from the density curve. Again, computations require more advanced mathematics.

The mean and variances of random variables obey the following rules. If a and b are fixed numbers then

$$\mu_{a + bX} = a + b\mu_X$$

$$\sigma^2_{a + bX} = b^2 \sigma^2_X$$

If X and Y are any two random variables, then

$$\mu_{X+Y} = \mu_X + \mu_Y$$

If X and Y are independent random variables, then

$$\sigma^2_{X+Y} = \sigma^2_X + \sigma^2_Y$$

$$\sigma^2_{X-Y} = \sigma^2_X + \sigma^2_Y$$

If X and Y have correlation ρ, then the general addition rule for variances of random variables is

$$\sigma^2_{X+Y} = \sigma^2_X + \sigma^2_Y + 2\rho\,\sigma_X\sigma_Y$$

$$\sigma^2_{X-Y} = \sigma^2_X + \sigma^2_Y - 2\rho\,\sigma_X\sigma_Y$$

GUIDED SOLUTIONS

Exercise 4.66

KEY CONCEPTS - independence, misconceptions about the law of large numbers

The key concept that must be properly understood to answer the questions raised in this problem is the notion of independence. Events A and B are independent if knowledge that A has occurred does not alter our assessment of the probability that B will occur. Do not be mislead by misconceptions based on a faulty understanding of the nature of random behavior or a faulty understanding of the law of large numbers (either that runs indicate that a hot streak is in progress and will continue for a while, or that a run of one type of outcome must be immediately balanced by a lack of the outcome for several trials).

Exercise 4.67

KEY CONCEPTS - rules for means and variances

a) Recall that if X and Y are any two random variables, then
$$\mu_{X+Y} = \mu_X + \mu_Y$$

In this problem, let X be the time to bring a part from a bin to its position on an automobile chassis and Y be the time required to attach the part to the chassis. Then the time for the total operation is $X + Y$. Use the information in the problem and the rule for the mean of $X + Y$ to evaluate the μ_{X+Y}, the mean time to complete the operation.

b) Do the variances affect the rules for means?

c) Does the correlation affect the rules for means?

Exercise 4.69

KEY CONCEPTS - rules for means and variances

Unlike the mean of $X + Y$, the variance of $X + Y$ depends on the correlation. Specifically the rule for variances is

$$\sigma^2_{X+Y} = \sigma^2_X + \sigma^2_Y + 2\rho\sigma_X\sigma_Y,$$

which simplifies to $\sigma^2_{X+Y} = \sigma^2_X + \sigma^2_Y$ when $\rho = 0$. Recall that when X and Y are independent, $\rho = 0$. Thus, you can use the simpler formula when X and Y are independent, while you need to use the general rule when $\rho = 0.3$.
Again, let X be the time to bring a part from a bin to its position on an automobile chassis and Y be the time required to attach the part to the chassis. Then the time for the total operation is $X + Y$. The problem asks you to find the standard deviation of the total time, so after finding the variance you must take the square root of your answer. You are given $\sigma_X = 2$ and $\sigma_Y = 4$.

i) $\rho = 0$ $\sigma^2_{X+Y} =$ $\sigma_{X+Y} =$

ii) $\rho = .3$ $\sigma^2_{X+Y} =$ $\sigma_{X+Y} =$

If you did the calculations correctly, the variance should be larger when the times are positively correlated. Try and explain in simple language why that should be the case.

Exercise 4.74

KEY CONCEPTS - rules for means and variances of random variables

a) The distribution of X is

Temperature (x)	540	545	550	555	560
Probability (p)	0.1	0.25	0.3	0.25	0.1

Recall that the average (mean) of X = earnings, is computed using the formula

$$\mu_X = x_1 p_1 + x_2 p_2 + \dots + x_k p_k$$

where the values of x_i and the p_i are given in the preceding table.

$$\mu_X =$$

Once you have calculated the mean μ_X, the general formula for the variance is

$$\sigma_X^2 = (x_1 - \mu_X)^2 p_1 + (x_2 - \mu_X)^2 p_2 + \ldots + (x_k - \mu_X)^2 p_k$$

After you have calculated the variance, remember to take the square root to obtain the standard deviation. If you are having trouble with these formulas, review Example 4.20 in the text.

$$\sigma_X^2 =$$

$$\sigma_X =$$

b) Calculations are relatively easy using the results of (a) if we recall the formulas

$$\mu_{a + bX} = a + b\mu_X$$

$$\sigma^2{}_{a + bX} = b^2 \sigma^2{}_X$$

What are a and b in this case?

c) This is just like (b). Again one must identify a and b.

Exercise 4.79

KEY CONCEPTS - mean of a random variable

Recall that the average (mean) of $X =$ earnings, is computed using the formula

$$\mu_X = x_1 p_1 + x_2 p_2 + \ldots + x_k p_k$$

where the values of x_i and the p_i are given in the following table. We have filled in the missing probability. How was it calculated?

Earnings (x)	-$99750	-$99500	-$99250	-$99000	-$98750	$1250
Probability (p)	0.00183	0.00186	0.00189	0.00191	0.00193	0.99058

Now use the formula to compute μ_X. The first five values of X are negative and the last is positive, so be careful with the signs of the products when computing the mean of X.

$$\mu_X =$$

COMPLETE SOLUTIONS

Exercise 4.66

a) Consecutive spins of a (fair) roulette wheel should be independent. Thus the particular results of previous spins will not change the probability of any particular outcome on the next spin. On the next spin black is just as likely as red. The gambler's reasoning that red is "hot" fails to recognize that spins are independent. It is based on a false understanding of random behavior.

b) The gambler is again wrong because he is assuming that consecutive cards are independent. This is not the case here. Initially, the deck contains 52 cards, half of which are red and half of which are black. However, each time a card is removed from the deck, the number of red and black cards remaining changes. For example, if I am dealt five red cards from a deck of 52 cards, the deck now contains only 47 cards of which 21 are red and 26 are black. The probability that the next card is red is 21/47 which is less than the probability that the next card is black, which is 26/47.

Exercise 4.67

a) We are told that $\mu_X = 11$ seconds and $\mu_Y = 20$ seconds. The mean time to complete the operation is

$$\mu_{X+Y} = \mu_X + \mu_Y = 11 + 20 = 31 \text{ seconds}$$

b) Changing the standard deviations does not change the rule for finding the mean of $X + Y$, so the mean time to complete the operation is still 31 seconds even if the standard deviations of each part of the operation are decreased.

c) The correlation affects the variance of $X + Y$, not its mean. The mean time to complete the operation is still 31 seconds even if the times required for the two steps are correlated.

Exercise 4.69

i) $\rho = 0$ $\sigma^2_{X+Y} = \sigma^2_X + \sigma^2_Y = (2)^2 + (4)^2 = 20$

   $\sigma_{X+Y} = \sqrt{20} = 4.47$ seconds

ii) $\rho = .3$ $\sigma^2_{X+Y} = \sigma^2_{X+Y} = \sigma^2_X + \sigma^2_Y + 2\rho\,\sigma_X\sigma_Y = (2)^2 + (4)^2 + (.3)(2)(4)$

   $= 22.4$

   $\sigma_{X+Y} = \sqrt{22.4} = 4.73$ seconds

When the times are positively correlated, if it takes longer than average time to bring the part from the bin to the chassis, it will also tend to take longer than average to attach the part making for a very long total time. Similarly, shorter than average times for bringing the part from the bin will be associated with shorter than average times to attach the part. This tends to make the total time further from the overall mean of 31 seconds in both directions, increasing the variability. If the times were independent, than a longer time to bring the part from the bin might have a shorter than average time to attach the part, and this cancellation brings the total time closer to the mean than when there is a positive correlation.

Exercise 4.74

a) We compute

   $\mu_X = 540(0.1) + 545(0.25) + 550(0.3) + 555(0.25) + 560(0.1)$

   $= 54 + 136.25 + 165 + 138.75 + 56$

   $= 550$

   $\sigma^2_X = (540 - 550)^2(0.1) + (545 - 550)^2(0.25) + (550 - 550)^2(0.25) +$
   $\qquad (555 - 550)^2(0.25) + (560 - 550)^2(0.1)$

   $= 10 + 6.25 + 0 + 6.25 + 10$

   $= 32.5$

   $\sigma_X = \sqrt{32.5} = 5.7$

b) Here we are interested in $X - 550$ so $a = -550$ and $b = 1$. Thus we have

   $\mu_{-550 + X} = -550 + 1\mu_X = -550 + 550 = 0$

   $\sigma^2_{-550 + X} = 1^2\sigma^2_X = 32.5$

hence

   $\sigma_{-550 + X} = \sqrt{32.5} = 5.7$

c) Here we want $(9/5)X + 32$ so $a = 32$ and $b = 9/5$. Thus

$$\mu_{32 + (9/5)X} = 32 + (9/5)\mu_X = 32 + (9/5)550 = 32 + 990 = 1022$$

$$\sigma^2_{32 + (9/5)X} = (9/5)^2\sigma^2_X = (81/25)32.5 = 105.3$$

hence

$$\sigma_{32 + (9/5)X} = \sqrt{105.3} = 10.26.$$

Exercise 4.79

We calculate,

$$\begin{aligned}
\mu_X &= 0.00183(-99750) + 0.00186(-99500) + 0.00189(-99250) \\
&\quad + 0.00191(-99000) + 0.00193(-98750) + 0.99058(1250) \\[6pt]
&= -182.5425 - 185.07 - 187.5825 - 189.09 - 190.5875 + 1238.225 \\[6pt]
&= 303.3525.
\end{aligned}$$

Not surprisingly, the mean is positive so the insurance company expects to make a little over \$303 per policy that it sells. Like any form of gambling, in the long run the insurance companies will make money despite an occassional large payout.

SECTION 4.5

OVERVIEW

This section discusses a number of basic concepts and rules that are used to calculate probabilities of complex events. The **complement** A^c of an event A contains all the outcomes in the sample space that are not in A. It is the "opposite" of A. The **union** of two events A and B contains all outcomes in A, in B, or in both. The union is sometimes referred to as the event A or B. The **intersection** of two events A and B contains all outcomes that are in both A and B simultaneously. The intersection is sometimes referred to as the event A and B. We say two events A and B are **disjoint** if they have no outcomes in common.

The **conditional probability** of an event B given an event A is denoted $P(B|A)$ and is defined by

$$P(B|A) = \frac{P(A \text{ and } B)}{P(A)}$$

when $P(A) > 0$. In practice it can often be determined directly from the information given in a problem. Two events A and B are **independent** if $P(B|A) = P(B)$.

Other general rules of elementary probability are

• Legitimate values: $0 \le P(A) \le 1$ for any event A

• Total probability: $P(S) = 1$, where S denotes the sample space.

• Complement rule: $P(A^c) = 1 - P(A)$

• Addition rule: $P(A \text{ or } B) = P(A) + P(B) - P(A \text{ and } B)$

• Multiplication rule: $P(A \text{ and } B) = P(A)P(B|A)$

• Bayes rule: $P(A|B) = \dfrac{P(B|A)P(A)}{P(B|A)P(A) + P(B|A^c)P(A^c)}$

$$\text{provided } 0 < P(A), P(B) < 1$$

• For disjoint events: $P(A \text{ and } B) = 0$ and so $P(A \text{ or } B) = P(A) + P(B)$

• For independent events: $P(A \text{ and } B) = P(A)P(B)$

In problems with several stages, it is helpful to draw a tree diagram to guide you in the use of the multiplication and addition rules.

GUIDED SOLUTIONS

Exercise 4.89

KEY CONCEPTS - the addition rule

This is an application of the addition rule $P(A \text{ or } B) = P(A) + P(B) - P(A \text{ and } B)$

Exercise 4.91

KEY CONCEPTS - Venn diagrams

a) From Exercise 4.89 we have the following facts about A and B.

$P(A) = 0.6$
$P(B) = 0.5$
$P(A \text{ and } B) = 0.3$

Below is a Venn diagram showing *A* and *B*. The event that Consolidated wins both jobs corresponds to {*A* and *B*}. Shade the portion representing the event {*A* and *B*}in the Venn diagram below.

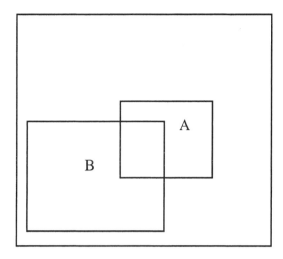

What is the probability that consolidated wins both jobs?

b) Below is a Venn diagram showing *A* and *B*. Shade the portion representing Consolidated wins the first job but not the second which is {*A* and B^c}.

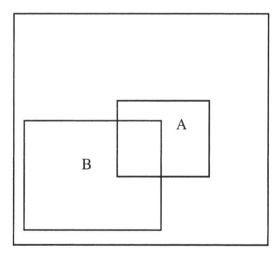

Use the facts listed in (a) and the Venn diagram to calculate *P(A* and B^c).

c) Below is a Venn diagram showing *A* and *B*. Shade the portion representing Consolidated does not win the first job but does the second. Write this event in terms of *A*, *B*, A^c and B^c.

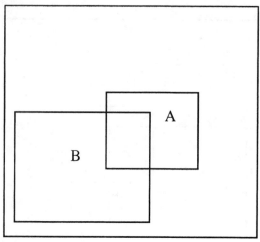

Use the facts listed in (a) and the Venn diagram to calculate the probability of this event.

d) Below is a Venn diagram showing *A* and *B*. Shade the portion representing Consolidated does not win either job. Write this event in terms of *A*, *B*, A^c and B^c.

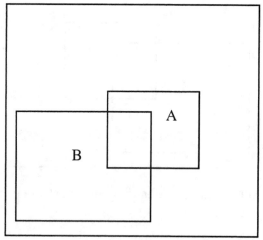

Use the facts listed in (a) and the Venn diagram to calculate the probability of this event.

Exercise 4.95

KEY CONCEPTS - conditional probabilities and the multiplication rule

The Table in Exercise 4.94 is reproduced below to assist you.

	Bachelor's	Master's	Professional	Doctorate	Total
Female	645	227	32	18	922
Male	505	161	40	26	732
Total	1150	388	72	44	1654

a) You can calculate this probability directly from the table. All degree recipients in the table are equally likely to be selected (that is what it means to select a degree recipient at random) so that the fraction of the degree recipients in the table that are men is the desired probability. How many degree recipients are men? Where do you find this in the table? What is the total number of degree recipients represented in the table? Use these numbers to compute the desired fraction.

b) This probability can also be calculated directly from the table. Since this is a conditional probability (i.e. this is a probability given that the degree recipient is a man), we restrict ourselves only to degree recipients that are men. The desired probability is then the fraction of these men that received a bachelor's degree. Use the appropriate entries in the table to compute this fraction.

c) Recall the multiplication rule says

$P($degree recipient is both "male" and "received a bachelor's"$)$

$= P($degree recipient is male$)P($degree recipient received bachelors $|$ degree recipient is male$)$

Using the answers in (a) and (b), evaluate this probability.

The number of degree recipients that are both "male" and "received a bachelor's" can be read directly from the table. What is this number? What fraction of the total number of degree recipients represented in the table is this number? This should agree with the probability you calculated using the multiplication rule.

Exercise 4.101

KEY CONCEPTS - multiplication rules and conditional probability

We are given the following probabilities.

$$P(\text{call not completed}) = 0.70$$

$$P(\text{talk to a man}) = 0.20$$

$$P(\text{talk to a woman}) = 0.10$$

$$P(\text{sale} \mid \text{talk to a woman}) = 0.3$$

$$P(\text{sale} \mid \text{talk to a man}) = 0.2$$

and

$$P(\text{sale} \mid \text{call not completed}) = 0.0$$

You are asked to compute the probability of a sale. Let A be the event "sale", B be the event "talk to a male," C the event "talk to a female," and D be the event "the call is not completed."

$$P(\text{sale}) = P(A) = P(A \text{ and } B) + P(A \text{ and } C) + P(A \text{ and } D)$$

First use the multiplication rule to evaluate $P(A \text{ and } B) = P(B)P(A \mid B)$, and similarly for $P(A \text{ and } C)$ and $P(A \text{ and } D)$, where we note that $P(A \text{ and } D) = 0$. Now combine your answers to evaluate

$$P(\text{sale}) =$$

Exercise 4.103

KEY CONCEPTS - Bayes rule

What we want to calculate is

P(female | sale) = *P*(female and sale) / *P*(sale).

This is an application of Bayes rule as we are calculating the "reverse" conditional probability from the ones we are given. Although it is an example of Bayes rule, it is better to work your way through the problem as in Examples 4.35 and 4.36 of the text rather than trying to memorize the rule.

P(female and sale) can be calculated using the multiplication rule and *P*(sale) has been computed in Exercise 4.101. Put these together to give the desired probability.

COMPLETE SOLUTIONS

Exercise 4.89

We are given that $P(A) = 0.6$, $P(B) = 0.5$, and $P(A \text{ and } B) = 0.3$, hence

$$P(A \text{ or } B) = 0.6 + 0.5 - 0.3 = 0.8$$

Exercise 4.91

a) The shaded area below is {*A* and *B*}.

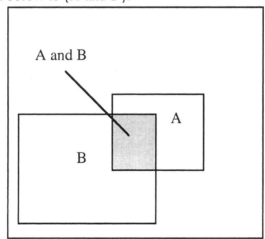

P(*A* and *B*) is given as 0.3.

b)The shaded area below is $\{A \text{ and } B^c\}$.

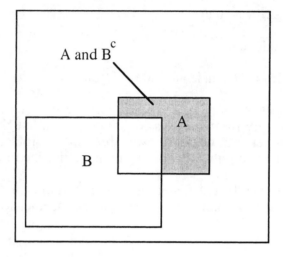

From the diagram we see that we can write

$$P(A \text{ and } B^c) = P(A) - P(A \text{ and } B) = 0.6 - 0.3 = 0.3$$

c) The shaded area below is $\{ A^c \text{ and } B \}$.

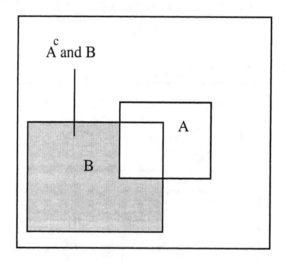

From the diagram we see that we can write

$$P(A^c \text{ and } B) = P(B) - P(A \text{ and } B) = 0.5 - 0.3 = 0.2$$

d) The shaded below area is { A^c and B^c }.

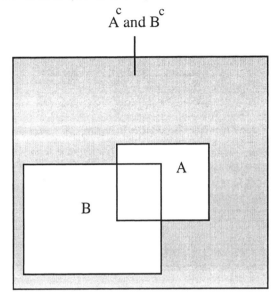

From the diagram we see that we can write

$$P(A^c \text{ and } B^c) = 1 - P(A \text{ or } B) = 1 - 0.8 = 0.2$$

where we have used the value of $P(A \text{ or } B) = 0.8$ calculated in problem 4.89.

Exercise 4.95

a) The number of degree recipients that are men is found in at the end of the row labeled "male" and is (in thousands) 732 The total number of degree recipients in the table is in the lower right corner and is (in thousands) 1654. The desired probability is thus

(number of degree recipients that are male)/(number of degree recipients)

$$= 732/1654$$

$$= 0.4426$$

b) The desired probability is

(number of males who received bachelors)/(number of males)

$$= 505/732$$

$$= 0.6899$$

c) Recall the multiplication rule says

P(degree recipient is both "male" and "received a bachelor's")

= *P*(degree recipient is male)*P*(degree recipient received bachelors | degree recipient is male)

= (0.4426)(0.6899) = 0.3053

The number of degree recipients that are both "male" and "received a bachelor's" can be read directly from the table and is (in thousands) 505 of the 1654 degree recipients, giving a probability of 505/1654 = 0.3053, which agrees with the answer obtained using the multiplication rule.

Exercise 4.101

KEY CONCEPTS - multiplication rules and conditional probability

P(*A* and *B*) = *P*(*B*)*P*(*A* | *B*) = (0.2)(0.2) = 0.04

P(*A* and *C*) = *P*(*C*)*P*(*A* | *C*) = (0.1)(0.3) = 0.03

and

P(sale) = 0.04 + 0.03 = 0.07

Exercise 4.103

P(female | sale) = *P*(female and sale) / *P*(sale)

P(female and sale) = *P*(sale | female)*P*(female) = (0.3)(0.1) = 0.03

and

P(sale) = 0.07 from Exercise 1.01, giving

P(female | sale) = 0.03/0.07 = 0.4286

CHAPTER 5

SAMPLING DISTRIBUTIONS

SECTION 5.1

OVERVIEW

One of the most common situations giving rise to a **count** X is the **binomial setting**. It consists of four assumptions about how the count was produced. They are

- the number n of observations is fixed
- the n observations are all independent
- each observation falls into one of two categories called "success" and "failure"
- the probability of success p is the same for each observation

When these assumptions are satisfied, the number of successes X has a **binomial distribution** with n trials and success probability p, denoted by $B(n, p)$. For smaller values of n, the probabilities for X can be found easily using statistical software. Table C in the text gives the probabilities for certain combinations of n and p, and there is also an exact formula. For large n, the **normal approximation** can be used.

For a large population containing a proportion p of successes, the binomial distribution is a good approximation to the number of successes in an SRS of size n, provided the population is at least 10 times larger than the sample. The mean and standard deviation for the binomial count X and the sample proportion $\hat{p} = X/n$ can be found using the formulas,

$$\mu_X = np \qquad\qquad \mu_{\hat{p}} = p$$

$$\sigma_X = \sqrt{np(1-p)} \qquad\qquad \sigma_{\hat{p}} = \sqrt{\frac{p(1-p)}{n}}$$

When n is large the count X is approximately $N(\,np,\ \sqrt{np(1-p)}\,)$ and the proportion $\hat{p}$ is approximately $N(\,p,\ \sqrt{\dfrac{p(1-p)}{n}}\,)$. These approximations should work well when $np \geq 10$ and $N(1-p) \geq 10$. For the count, the **continuity correction** improves the approximation, particularly when the values of n and p are closer to these cutoffs.

The exact **binomial probability formula** is given by

$$P(X=k) = \binom{n}{k} p^k (1-p)^{n-k}$$

where $k = 0, 1, 2, ..., n$ and $\binom{n}{k}$ is the **binomial coefficient**

$$\binom{n}{k} = \frac{n!}{k!(n-k)1}$$

and the factorial $n!$ is

$$n! = n \times (n-1) \times (n-2) \times \cdots \times 3 \times 2 \times 1$$

GUIDED SOLUTIONS

Exercise 5.1

KEY CONCEPTS - binomial setting

There are four assumptions that need to be satisfied to ensure that the count X has a binomial distribution. The number of observations or trials is fixed in advance, each trial results in one of two outcomes, the trials are independent and the probability of success is the same from trial to trial. In addition, for a large population with a proportion p of successes, we can use the binomial distribution as an approximation to the distribution of the count X of successes in an SRS of size n. For each setting below, see if all four assumptions are satisfied.

a) Think about how this fits in the binomial setting. What is n? What are the two outcomes, and why might the trials be considered independent?

b) Think about how many observations or trials there are going to be.

c) Think about whether the trials are independent.

Exercise 5.5

KEY CONCEPTS - binomial probabilities

a) Suppose we let X denote the number of members of the committee that are Hispanic. We are in the binomial setting with $n = 15$ trials, and letting "success" correspond to being Hispanic, we have $p = 0.3$. We want to compute the probability that *exactly* 3 members of the committee are Hispanic, namely we want $P(X = 3)$. Table C can be used to compute this probability. Find the column labelled by 0.30 and look for the row corresponding to $n = 15$ and $k = 3$. Alternatively, you can use statistical software to compute this probability. What do you find?

$P(X = 3) =$

b) Now we want the probability that 3 or fewer members of the committee are Hispanic. Express this probability in terms of X and add the appropriate entries in Table C to compute this probability. Alternatively, you may wish to use statistical software to compute the probability.

Exercise 5.11

KEY CONCEPTS - mean of the binomial, normal approximation for counts

a) Let X denote the number of home runs Mark McGwire will hit in 509 times at bat. The problem tells us to treat X as a count having the $B(509, 0.116)$ distribution. We are asked to compute the mean μ_X of X. What is the formula for μ_X? To answer this, see the Chapter Overview above, or the Summary of Section 5.1 in the text. Now use this formula to compute the mean.

$\mu_X =$

b) Now we are asked to compute $P(X \geq 70)$. This can be done using the normal approximation for a count. First check to see that np and $n(1-p)$ are both greater than or equal to 10. Next we need the mean and standard deviation of X. You computed the mean in a). To compute the standard deviation σ_X, use the formula given in the Chaper Overview on the previous page, or the Summary of Section 5.1 in the text. Write the result below.

$\sigma_X =$

Next use the the normal approximation to evaluate $P(X \geq 70)$. This is a normal probability calculation like those discussed in Section 3 of Chapter 1. You may want to review the material there to refresh your memory on how to do such calculations. The first step is to standardize the number 70 (compute its z-score) by subtracting the mean μ_X and dividing the result by σ_X. Next, use Table A to determine the area to the right of this z-score. It may be helpful to draw a normal curve to visualize the area.

$P(X \geq 70) =$

c) To answer this question, we now let X denote the number of home runs Barry Bonds will hit in 476 times at bat. The problem tells us to treat X as a count having the $B(476, 0.0865)$ distribution. We are asked to compute $P(X \geq 73)$. To do so you will again need to use the normal approximation. First compute

$\mu_X =$

$\sigma_X =$

and then use the same method as in (b) to complete the calculation.

$P(X \geq 73) =$

Exercise 5.13

KEY CONCEPTS - proportions, normal approximation for proportions

a) Let X denote the number of students in the SRS of $n = 100$ that support a crackdown on underage drinking. In this case, $X = 62$. To compute the sample proportion use the formula
sample proportion $= \hat{p} = X/n =$

b) If the proportion of all students on your campus who support a crackdown is $p = 0.67$, then X is a count having the $B(100, 0.67)$ distribution. You need to compute a probability about $\hat{p}$, namely $P(\hat{p} \le 0.62)$, because 0.62 is the sample proportion that resulted from the administration's sample. Since $\hat{p} \le 0.62$ whenever $X \le 62$, if you have statisticial software which computes binomial probabilities, try to find the exact probability.

If you don't have access to statistical software that computes binomial probabilities, you'll need to use the normal approximation to the sampling distribution of $\hat{p}$ to approximate the probability. Since $np = 100(0.67) = 67$ and $n(1 - p) = 100(0.33) = 33$ are both greater than or equal to 10, we can use the approximation. To use the normal approximation, the mean and standard deviation of $\hat{p}$ must be computed. Do this using the formulas given below.

mean $\qquad\qquad\qquad \mu_{\hat{p}} = p =$

standard deviation $\qquad\qquad \sigma_{\hat{p}} = \sqrt{\dfrac{p(1-p)}{n}} =$

Computing $P(\hat{p} \le 0.62)$ is a normal probability calculation like those in Section 3 of Chapter 1. You may want to review the material there in order to refresh your memory on how to do such calculations, as they form the basis of solving many of the exercises in this chapter. To compute the probability, we standardize the number 0.62 (compute its z-score) by subtracting the mean $\mu_{\hat{p}}$ and dividing the result by $\sigma_{\hat{p}}$. Next use Table A to determine the area to the left of this z-score. It may be helpful to draw a normal curve and the desired area to help you solve byvisualizing the area.

$P(\hat{p} \le 0.62) =$

c) You should base your comments on the probability calculated in (b). In particular, is this probability large enough that a value as small as 0.62 could plausibly arise by chance if the actual value is 0.67?

Exercise 5.22

KEY CONCEPTS - binomial probabilities, mean and variance of the binomial

a) Let $X =$ number of truthful persons "failing" the lie detector test. X has a binomial distribution with $n = 12$ and $p = 0.2$. The binomial probabilities can be found in Table C in your text or using statistical software. The probabilities below were obtained using statistical software.

k	$P(X = k)$
0	0.068719
1	0.206158
2	0.283468
3	0.236223
4	0.132876
5	0.053150
6	0.015502
7	0.003322
8	0.000519
9	0.000058
10	0.000004
11	0.000000
12	0.000000

Use these (or those in Table C) to evaluate the probability required.

b) Use the formula which expresses the mean and standard deviation of the binomial in terms of n and p.

mean =
standard deviation =

c) Don't be confused by the fact that the mean is not a whole number. Just determine which values of X are less than the mean and then add up their probabilities.

Exercise 5.23

KEY CONCEPTS - normal approximation for counts

a) The population proportion is

$$p = \frac{\text{number of blacks in the 2000 census}}{\text{total number of adults}}$$

Use the numbers given in the statement of the problem to compute p.

$p =$

b) Let X denote the number of blacks in the sample. X is a count and if the sample is a random sample, X will have a binomial distribution. What are n and

p in this case? Now use the formulas which express the mean of X in terms of n and p.

mean =

c) This part can be done using the normal approximation for a count. First check to see that np and $n(1-p)$ are both greater than or equal to 10. Next compute the standard deviation of X using the formula that expresses the standard deviation in terms of n and p.

standard deviation =

Now use the mean from (b), this standard deviation, and the normal approximation to evaluate $P(X \leq 170)$.

$P(X \leq 170) =$

COMPLETE SOLUTIONS

Exercise 5.1

a) The number of trials 50 is fixed, each child is a boy or girl, whether or not a child is a boy will not alter the probability that any other children are either girls or boys (excluding identical twins to keep it simple), and the probability of any child being a boy should be the same (about 0.5) for each birth. The binomial distribution should be a good probability model for the number of girls.

b) Although each birth is a boy or girl, we are not counting the number of successes in a fixed number of births. The number of observations (births) is random. The assumption of a fixed number of observations is violated.

c) This is not the same as interviewing 50 people at random. The husband and wife may tend to share the same opinion, and the trials will not be independent of each other. If we know the wife agrees, then it is more likely that the husband agrees than if we didn't have this information. The assumption of independent observations is violated.

Exercise 5.5

a) From Table C we find $P(X = 3) = .1700$.

b) We want

$$P(X \leq 3) = P(X = 0) + P(X = 1) + P(X = 2) + P(X = 3)$$

Using Table C to find each of the probabilities on the right hand side of this equation we find

$$P(X \leq 3) = .0047 + .0305 + .0916 + .1700 = .2968$$

Exercise 5.11

a) $\mu_X = np = 509 \times 0.116 = 59.044$.

b) First we check that

$np = 509 \times 0.116 = 59.044 \geq 10$, and $n(1\text{-}p) = 509 \times 0.884 = 449.956 \geq 10$

Next we compute

$$\mu_X = np = 509 \times 0.116 = 59.044$$

$$\sigma_X = \sqrt{np(1-p)} = \sqrt{509 \times 0.116 \times 0.884} = \sqrt{52.195} = 7.225$$

When n is large, X is approximately $N(np, \sqrt{np(1-p)}) = N(59.044, 7.225)$. The z-score of 70 is thus

$$z\text{-score of } 70 = \frac{70 - 59.044}{7.225} = 1.52$$

and so using Table A,

$$P(X \geq 70) = P(z \geq 1.52) = 1 - P(z \leq 1.52) = 1 - 0.9357 = 0.0643.$$

c) We compute

$$\mu_X = np = 476 \times 0.0865 = 41.174$$

$$\sigma_X = \sqrt{np(1-p)} = \sqrt{476 \times 0.0865 \times 0.9135} = \sqrt{37.612} = 6.133$$

and so the z-score of 73 is

$$z\text{-score of } 73 = \frac{73 - 41.174}{6.133} = 5.19$$

and so using Table A,

$$P(X \geq 73) = P(z \geq 5.19) = 1 - P(z \leq 5.19) \leq 1 - 0.9999 = 0.0001$$

Exercise 5.13

a) Sample proportion $= \hat{p} = X/n = \frac{62}{100} = 0.62$.

b) Using statistical software we find the exact probability of $X \leq 62$ to be 0.1690. To use the normal approximation we compute

$$\text{mean} = \mu_{\hat{p}} = p = 0.67$$

$$\text{standard deviation} = \sigma_{\hat{p}} = \sqrt{\frac{p(1-p)}{n}} = \sqrt{\frac{0.67 \times 0.33}{100}} = \sqrt{0.0022} = 0.047$$

From this, we compute the z-score of 0.62 to be

$$z\text{-score} = \frac{0.62 - 0.67}{0.047} = -1.06$$

Using Table A we find

$$P(\hat{p} \leq 0.62) = P(z \leq -1.06) = 0.1446$$

(Note: This differs somewhat from the exact value obtained using statistical software. If you use the continuity correction, you get $P(\hat{p} \leq 0.62) = 0.1685$.)

c) It is true that the sample proportion of 0.62 is smaller than the national value of 0.67. However, our probability calculation shows that a sample value of 0.62 (or smaller) has a probability 0.1446 of occurring simply by chance if the actual proportion on campus is 0.67. In most scientific journals this probability would be considered too large to allow one to assert with any degree of confidence that the survey supports the statement "that support for a crackdown is lower on our campus than nationally."

Exercise 5.22

a) X is B(12, 0.2). You are asked to evaluate $P(X \geq 1)$, the probability that the polygraph says that at least one person telling the truth is deceptive. Using the table of probabilities

k	$P(X = k)$
0	0.068719
1	0.206158
2	0.283468
3	0.236223
4	0.132876
5	0.053150
6	0.015502
7	0.003322
8	0.000519
9	0.000058
10	0.000004
11	0.000000
12	0.000000

you may at first think to add the probabilites for $k = 1, 2, 3, ..., 12$. While this will give the correct answer, in this case it is much simpler to use the rule for complements. $P(X \geq 1) = 1 - P(X = 0) = 1 - 0.068719 = 0.931281$. When determing the complementary event, you must be careful whether an event is of the form greater than or greater than or equal to. The complement of $X \geq 2$ is $X \leq 1$, while the complement of $X > 2$ is $X \leq 2$.

b) The mean is $np = 12(0.2) = 2.4$, and the standard deviation is found using the formula $\sqrt{np(1-p)} = \sqrt{12(0.2)(0.8)} = 1.386$.

c) X is less than the mean 2.4 only if X is 0, 1 or 2. So, we need to find
$$P(X \leq 2) = 0.068719 + 0.206158 + 0.283468 = 0.558345$$

Exercise 5.23

a) $p = \dfrac{23,772,494}{209,128,094} = 0.1137$.

b) X has a binomial distribution with $n = 1500$ and $p = 0.1137$. The mean of X is
$$\text{mean} = np = 1500(0.1137) = 170.55.$$

c) The normal approximation can be used since both $np = 170.55$ and $n(1-p) = 1329.45$ are greater than 10. We compute
$$\text{standard deviation} = \sqrt{np(1-p)} = \sqrt{1500(0.1137)(0.8863)} = 12.29$$

Using the mean in (b) and this standard deviation gives the approximation
$$P(X \leq 170) = P\left(\frac{X - 170.55}{12.29} \leq \frac{170 - 170.55}{12.29} \right) = P(z \leq -0.04) = 0.4840$$

SECTION 5.2

OVERVIEW

This section examines properties of the **sample mean** $\bar{x}$. If we select an SRS of size n from a large population with mean μ and standard deviation σ, the sample mean $\bar{x}$ has a sampling distribution with
$$\text{mean} = \mu_{\bar{x}} = \mu$$

and

$$\text{standard deviation} = \sigma_{\bar{x}} = \frac{\sigma}{\sqrt{n}}$$

This implies that the sample mean is an unbiased estimator of the population mean and is less variable than a single observation.

Linear combinations (such as sums or means) of independent normal random variables have normal distributions. In particular, if the population has a normal distribution, the sampling distribution of $\bar{x}$ is normal. Even if the population does not have a normal distribution, for large sample sizes the sampling distribution of $\bar{x}$ computed from an SRS is approximately normal. In particular, the **central limit theorem** states that for large n, the sampling distribution of $\bar{x}$ computed from an SRS is approximately $N(\mu, \frac{\sigma}{\sqrt{n}})$ for any population with mean μ and finite standard deviation σ.

SAMPLE PROBLEMS

GUIDED SOLUTIONS

EXERCISE 5.33

KEY CONCEPTS - the sampling distribution of the sample mean, normal probability calculations.

a) We take X to be the (numeric) grade of a randomly chosen student. The probability distribution of X is seen to be

Grade (X)	4	3	2	1	0
Probability	0.18	0.32	0.34	0.09	0.07

Refer to Section 4.4 for the formulas for the mean and standard deviation of a discrete random variable. Use these to compute the following.

μ = mean of X=

σ = standard deviation of X =

b) Recall that if we select an SRS of size n from a large population with mean μ and standard deviation σ, the sample mean $\bar{x}$ has a sampling distribution with

mean of $\bar{x} = \mu_{\bar{x}} = \mu =$

and

standard deviation of $\bar{x} = \sigma_{\bar{x}} = \dfrac{\sigma}{\sqrt{n}} =$

In this case, what is n, the sample size?

c) You can compute the probability $P(X \geq 3)$ that a randomly chosen Accounting 201 student gets a grade of B or better directly using the probabilility distribution in part (a). This is just like the calculation in Section 4.3.

$P(X \geq 3) =$

To compute the approximate probability $P(\bar{x} \geq 3)$ that the grade point average for 50 randomly chosen Accounting 201 students is B or better, we use the central limit theorem that states that for large n, the sampling distribution of $\bar{x}$ computed from an SRS is approximately $N(\mu, \dfrac{\sigma}{\sqrt{n}})$ for any population with mean μ and finite standard deviation σ. Thus, calculating $P(\bar{x} \geq 3)$ is just a normal probability calculation, like those we did in Section 3 of Chapter 1. You may wish to review the material there in order to refresh your memory as to how to do such calculations. We want the probability that the sample mean $\bar{x}$ is 3 or higher. To compute this probability, we standardize the number 3 (compute its z-score) by subtracting the mean $\mu_{\bar{x}}$ and dividing the result by $\sigma_{\bar{x}}$. You computed $\mu_{\bar{x}}$ and $\sigma_{\bar{x}}$ in (b). Next use Table A to determine the area to the right of this z-score under a standard normal curve. You may wish to draw a picture of the standard normal to help you visualize the desired area.

$P(\bar{x} \geq 3) =$

EXERCISE 5.36

KEY CONCEPTS - means and standard deviations of random variables, the law of large numbers, the central limit theorem, the sampling distribution of the sample mean, normal probability calculations

a) You may wish to review how to calculate the mean and standard deviation of a random variable. This was discussed in Chapter 4. Recall that if X is a discrete random variable having possible values $x_1, x_2, \ldots, x_k$ with corresponding probabilities $p_1, p_2, \ldots, p_k$, the mean μ_X is the average of the possible values weighted by the corresponding probabilities, i.e.

$$\mu_X = x_1 p_1 + x_2 p_2 + \ldots + x_k p_k$$

the variance is

$$\sigma^2_X = (x_1 - \mu_X)^2 p_1 + (x_2 - \mu_X)^2 p_2 + \ldots + (x_k - \mu_X)^2 p_k$$

and the standard deviation σ_X is the positive square root of the variance.

For simplicity, assume that the gambler bets on red. What is the probability he will win, assuming all slots are equally likely to contain the ball? Complete the table below to help you compute the mean and standard deviation of the outcomes.

Outcome (winnings) X	$1	-$1
Probability		

$\mu_X =$

$\sigma^2_X =$

$\sigma_X =$

b) Recall that the law of large numbers (see Chapter 4) tells us that the average of the values of X observed in many trials must approach μ_X. Interpret this in the context of this problem, using plain English.

c) The central limit theorem states that for large n, the sampling distribution of the mean winnings computed from an SRS is approximately $N(\mu_X, \frac{\sigma_X}{\sqrt{n}})$. What are n, μ_X, and σ_X here?

Refer to Chapter 1 to refresh your memory concerning the 68-95-99.7 rule. Remember to use μ_X for the mean and $\frac{\sigma_X}{\sqrt{n}}$ for the standard deviation when applying the rule.

d) From (c) we found that the sampling distribution of the gambler's mean winnings $\bar{x}$ if he makes 50 bets is $N(\mu_X = -0.0526, \frac{\sigma_X}{\sqrt{n}} = 0.1412)$. We want to compute the probability that $\bar{x}$ is less than 0. This is just a normal probability calculation of the sort we studied in Section 3 of Chapter 1. Find the z-score of 0 and then use Table A to determine the area under the standard normal curve below this z-score.

e) Now repeat the same sort of calculations as in (c) and (d), but with $n = 100,000$ rather than $n = 50$.

EXERCISE 5.43

KEY CONCEPTS - the sampling distribution of the sample mean, normal probability calculations

We are told that the distribution of individual scores at Southwark Elementary School is approximately normal with mean $\mu = 13.6$ and standard deviation $\sigma = 3.1$. To find L, Mr. Lavin needs to first determine the sampling distribution of the mean score $\bar{x}$ of $n = 22$ children. What is this sampling distribution?

To complete the problem, you need to find L such that the probability of $\bar{x}$ being below L is only 0.05. We did this sort of problem in Section 3 of Chapter 1. First, refer to Table A to find the value z such that the area to the left of z under a standard normal curve is 0.05. What is this value?

This value z is the z-score of L. This means $z = (L - \mu_{\bar{x}})/\sigma_{\bar{x}}$, where $\mu_{\bar{x}}$ and $\sigma_{\bar{x}}$ are the mean and standard deviation, respectively, of the sampling distribution of $\bar{x}$. Solve this equation for L.

EXERCISE 5.45

KEY CONCEPTS - linear combinations of independent random variables

a) We need to recall several facts from Section 4 of Chapter 4. First, recall that for a random variable X

$$\mu_{a + bX} = a + b\mu_X$$

Using $a = 0$ and $b = -1$, this implies that $\mu_{-X} = -\mu_X$. Second, recall that for random variables X and Y

$$\mu_{Y+X} = \mu_Y + \mu_X$$

Replacing X by the random variable $-X$ and using $\mu_{-X} = -\mu_X$ we have

$$\mu_{Y-X} = \mu_Y - \mu_X$$

Finally, recall that if X and Y are independent random variables, then

$$\sigma^2_{Y-X} = \sigma^2_Y + \sigma^2_X.$$

We now apply these facts to $\bar{x}$ and $\bar{y}$ to get

$$\mu_{\bar{y}-\bar{x}} = \mu_{\bar{y}} - \mu_{\bar{x}}$$

and

$$\sigma^2_{\bar{y}-\bar{x}} = \sigma^2_{\bar{y}} + \sigma^2_{\bar{x}}$$

We take the square root of $\sigma^2_{\bar{y}-\bar{x}}$ to get the standard deviation of $\bar{y} - \bar{x}$. To complete the problem, what are $\mu_{\bar{y}}$, $\mu_{\bar{x}}$, $\sigma^2_{\bar{y}}$, and $\sigma^2_{\bar{x}}$ here?

b) We know that since weight gains are normally distributed in both populations, $\bar{x}$ and $\bar{y}$ are normally distributed. You determined the mean and standard deviations of the sampling distributions of these means in (a). Since we can assume $\bar{x}$ and $\bar{y}$ are independent, we can assume the sampling distribution of $\bar{y} - \bar{x}$ is also normal. You determined the mean and standard deviation of the sampling distribution of $\bar{y} - \bar{x}$ in (a).

c) Here we want to determine the probability that $\bar{y} - \bar{x}$ is greater than 25. We found that the sampling distribution of $\bar{y} - \bar{x}$ is $N(\mu_{\bar{y}-\bar{x}} = 25, \sigma_{\bar{y}-\bar{x}} = 16.62)$ in (b). The desired probability can be computed using the methods discussed in Section 3 of Chapter 1 for calculating normal probabilities.

COMPLETE SOLUTIONS

EXERCISE 5.33

a) μ = mean of X = $4 \times 0.18 + 3 \times 0.32 + 2 \times 0.34 + 1 \times 0.09 + 0 \times 0.07 = 2.45$

To compute the standard deviation we first compute the variance

$\sigma^2 = (4\text{-}2.45)^2 \times 0.18 + (3\text{-}2.45)^2 \times 0.32 + (2\text{-}2.45)^2 \times 0.34 + (1\text{-}2.45)^2 \times 0.09$

$\qquad + (0\text{-}2.45)^2 \times 0.07 = 1.2075$

and so

$$\sigma = \text{standard deviation of } X = \sqrt{1.2075} = 1.099$$

b) Using the results in part (a) and the fact that the sample size is $n = 50$ we obtain

$$\text{mean of } \bar{x} = \mu_{\bar{x}} = \mu = 2.45$$

$$\text{standard deviation of } \bar{x} = \sigma_{\bar{x}} = \frac{\sigma}{\sqrt{n}} = \frac{1.099}{\sqrt{50}} = 0.155$$

c) $P(X \geq 3) = P(X = 3) + P(X = 4) = 0.32 + 0.18 = 0.50.$

Using the results of (b), the sampling distribution of $\bar{x}$ is approximately $N(2.45, 0.155)$. Thus

$$P(\bar{x} \geq 3) = P\left(\frac{\bar{x} - 2.45}{0.155} \geq \frac{3 - 2.45}{0.155} \right) = P(z \geq 3.55) = 1 - P(z \leq 3.55)$$

The value 3.55 is outside the range of Table A, and all we can say is that $P(z \leq 3.55)$ is larger than 0.9998, hence $P(\bar{x} \geq 3)$ is smaller than $1 - 0.9998 = 0.0002$.

EXERCISE 5.36

a) Since 18 of the 38 slots are red, the probability of winning (if you bet on red) is 18/38. Thus we have

Outcome (winnings) X	$1	-$1
Probability	18/38	20/38

and so

$$\text{mean winnings} = \mu_X = (\$1)(18/38) + (-\$1)(20/38) = -\$2/38 = -\$0.0526$$

$$\begin{aligned}\text{variance of winnings (in dollars squared)} &= \sigma^2_X \\ &= (1 - [-2/38])^2(18/38) + (-1 - [-2/38])^2(20/38) \\ &= 0.5249 + 0.4724 \\ &= 0.9973\end{aligned}$$

$$\text{standard deviation of winnings} = \sigma_X = \$\sqrt{0.9973} = \$0.9986$$

b) The law of large numbers tells us that if a gambler makes a large number of bets on red, his mean winnings per bet will be approximately $\mu_X = -\$0.0526$. In other words, he will lose \$0.0526 on the average per bet.

c) Here $n = 50$, $\mu_X = -\$0.0526$, and $\sigma_X = \$0.9986$. The approximate distribution of the gambler's mean winnings in 50 bets is thus normal with

$$\text{mean} = \mu_X = -\$0.0526$$

and

$$\text{standard deviation} = \frac{\sigma_X}{\sqrt{n}} = \frac{\$0.9986}{\sqrt{50}} = \$0.1412$$

The 68-95-99.7 rule says that the middle 95% of the gambler's mean winnings on nights when he places 50 bets is within two standard deviations of the mean. In other words, it lies between $\mu_X \pm 2\frac{\sigma_X}{\sqrt{n}} = -\$0.0526 \pm 2(\$0.1412)$ or between $-\$0.3350$ and $\$0.2298$. Multiplying by 50 to convert to total winnings, we get that the middle 95% of the gambler's total winnings on nights when he places 50 bets is between $-\$16.75$ and $\$11.49$.

d) The sampling distribution of the gambler's mean winnings $\bar{x}$ if he makes 50 bets is $N(\mu_X = -0.0526, \frac{\sigma_X}{\sqrt{n}} = 0.1412)$. To find the probability that $\bar{x}$ is less than 0, we compute the z-score of 0, which is

$$z\text{-score} = (0 - [-0.0526])/(0.1412) = 0.0526/0.1412 = 0.37$$

According to Table A, the area under the standard normal curve to the left of 0.37 is 0.6443. Thus the probability that the gambler will lose money if he makes 50 bets is 0.6443.

e) If $n = 100,000$ bets are made on red, the sampling distribution of the mean winnings is again normal but now with

$$\text{mean} = \mu_X = -\$0.0526$$

and

$$\text{standard deviation} = \frac{\sigma_X}{\sqrt{n}} = \frac{\$0.9986}{\sqrt{100,000}} = \$0.00316$$

The 68-95-99.7 rule says that the middle 95% of the mean winnings of gamblers on these 100,000 bets is within two standard deviations of the mean.

In other words, lies between $\mu_X \pm 2\dfrac{\sigma_X}{\sqrt{n}} = $ -\$0.0526 $\pm$ 2(\$0.00316) or between -\$0.05892 and -\$0.04628. Multiplying by 100,000 to convert to total winnings, we get that the middle 95% of the mean winnings on 100,000 bets is between -\$5892 and -\$4628.

EXERCISE 5.43

The sampling distribution of the mean score $\bar{x}$ of 22 children is approximately normal with mean

$$\mu_{\bar{x}} = \mu = 13.6$$

and standard deviation

$$\sigma_{\bar{x}} = \frac{\sigma}{\sqrt{n}} = \frac{3.1}{\sqrt{22}} = 0.66$$

Next, we note that from Table A the value of z such that the area to the left of it under a standard normal curve is 0.05 is $z = -1.65$. Thus

$$-1.65 = (L - 13.6)/0.66$$

Solving for L gives

$$L = (-1.65)(0.66) + 13.6 = 12.51$$

EXERCISE 5.45

a) We know that the mean and standard deviation of X are 360g and 55g and the mean and standard deviation of Y are 385g and 50g. For an average based on a sample size of $n = 20$, we have

$$\mu_{\bar{x}} = 360$$

$$\sigma_{\bar{x}}^2 = (55)^2/(20) = 151.25$$

$$\mu_{\bar{y}} = 385$$

$$\sigma_{\bar{y}}^2 = (50)^2/(20) = 125$$

Therefore

$$\mu_{\bar{y}-\bar{x}} = \mu_{\bar{y}} - \mu_{\bar{x}} = 385 - 360 = 25$$

$$\sigma^2_{\bar{y}-\bar{x}} = \sigma^2_{\bar{y}} + \sigma^2_{\bar{x}} = 151.25 + 125 = 276.25$$

$$\sigma_{\bar{y}-\bar{x}} = \sqrt{276.25} = 16.62.$$

b) From (a) we have

$$\mu_{\bar{x}} = 360$$

$$\sigma^2_{\bar{x}} = (55)^2/(20) = 151.25 \text{ (hence } \sigma_{\bar{x}} = \sqrt{151.25} = 12.30)$$

$$\mu_{\bar{y}} = 385$$

$$\sigma^2_{\bar{y}} = (50)^2/(20) = 125 \text{ (hence } \sigma_{\bar{y}} = \sqrt{125} = 11.18)$$

$$\mu_{\bar{y}-\bar{x}} = 25$$

$$\sigma_{\bar{y}-\bar{x}} = 16.62.$$

Thus,

the distribution of $\bar{x}$ is $N(\mu_{\bar{x}} = 360, \ \sigma_{\bar{x}} = 12.30)$

the distribution of $\bar{y}$ is $N(\mu_{\bar{y}} = 385, \ \sigma_{\bar{y}} = 11.18)$

the distribution of $\bar{y} - \bar{x}$ is $N(\mu_{\bar{y}-\bar{x}} = 25, \ \sigma_{\bar{y}-\bar{x}} = 16.62)$.

c) To determine the probability that $\bar{y} - \bar{x}$ is greater than 25, we compute the z-score of 25. We get

$$z\text{-score} = (25 - \mu_{\bar{y}-\bar{x}})/\sigma_{\bar{y}-\bar{x}} = (25 - 25)/16.62 = 0.$$

The area to the right of 0 under a standard normal curve is 0.5, so this is the probability that $\bar{y} - \bar{x}$ is greater than 25.

CHAPTER 6

INTRODUCTION TO INFERENCE

SECTION 6.1

OVERVIEW

A **confidence interval** provides an estimate of an unknown parameter of a population or process along with an indication of how accurate this estimate is and how **confident** we are that the interval is correct. Confidence intervals have two parts. One is an interval computed from our data. This interval typically has the form

$$\text{estimate} \pm \text{margin of error}$$

The other part is the **confidence level**, which states the probability that the <u>method</u> used to construct the interval will give a correct answer. For example, if you use a 95% confidence interval repeatedly, in the long run 95% of the intervals you construct will contain the correct parameter value. Of course, when you apply the method only once you do not know if your interval gives a correct value or not. Confidence refers to the probability that the method gives a correct answer in repeated use, not the correctness of any particular interval we compute from data.

Suppose we wish to estimate the unknown mean μ of a normal population with known standard deviation σ based on an SRS of size n. A level C confidence interval for μ is

$$\bar{x} \pm z^* \frac{\sigma}{\sqrt{n}}$$

where z^* is such that the probability is C that a standard normal random variable lies between $-z^*$ and z^* and is obtained from the bottom row in Table D.

The margin of error $z^* \dfrac{\sigma}{\sqrt{n}}$ of a confidence interval decreases when any of the following occur:

• the confidence level C decreases

• the sample size n increases

• the population standard deviation σ decreases.

The sample size needed to obtain a confidence interval for a normal mean of the form

$$\text{estimate} \pm \text{margin of error}$$

with a specified margin of error m is

$$n = \left(\frac{z^* \sigma}{m} \right)^2$$

where z^* is the critical point for the desired level of confidence. Many times the n you will find will not be an integer. If it is not, round up to the next larger integer.

The formula for any specific confidence interval is a recipe that is correct under specific conditions. The most important conditions concern the methods used to produce the data. Many methods (including those discussed in this section) assume that our data were collected by random sampling. Other conditions, such as the actual distribution of the population, are also important.

SAMPLE PROBLEMS

GUIDED SOLUTIONS

Exercise 6.5

KEY CONCEPTS - the sampling distribution of $\bar{x}$, confidence intervals

a) We are told that the $n = 100$ invoices are a random sample, and we know that the sampling distribution of $\bar{x}$ has standard deviation $\sigma_{\bar{x}} = \dfrac{\sigma}{\sqrt{n}}$. Use the values of σ and n to compute $\sigma_{\bar{x}}$.

$\sigma_{\bar{x}} =$

b) Recall from Section 3 of Chapter 1, the 68 - 95 - 99.7 rule, says that 68% of the area under a normal curve lies within one standard deviation of the mean, 95% within two standard deviations of the mean, and 99.7% within three standard deviations of the mean. Now you should be able to fill in the blank.

c) Use the value of $\sigma_{\bar{x}}$ that you computed in part (a) and your answer to part (b) to fill in the blank.

Exercise 6.19

KEY CONCEPTS - confidence intervals, the sample size required to obtain a confidence interval of specified margin of error

a) The margin of error of a level C confidence interval is $z^* \dfrac{\sigma}{\sqrt{n}}$, where z^* is such that the probability is C that a standard normal random variable lies between $-z^*$ and z^* and is obtained from the bottom row in Table D. To do this exercise, you must identify

   C = the level of confidence required =

   z^* = the probability is C that a standard normal random variable lies
        between $-z^*$ and z^* =

   σ = population standard deviation =

   n = the sample size used =

Determine the above values and then compute the margin of error

$$m = z^* \frac{\sigma}{\sqrt{n}} =$$

b) The calculations here are the same as in (a) except that n has changed from 100 to 10. Should the margin of error be larger or smaller? Why? Repeat the calculations you did in (a) with this change and see if your intuition was correct.

c) The smallest value of n that will yield a 95% confidence interval with a margin of error $= m$ must satisfy

$$n = \left(\frac{z^* \sigma}{m} \right)^2$$

where z^* and σ are as in (a). Identify m in this case and use the above formula to compute n. Remember to round your answer up to the nearest integer. Is n within the limits of your budget as indicated in (a)?

Exercise 6.26

KEY CONCEPTS - confidence levels for several confidence intervals simultaneously, binomial probability calculations

We know the following:

- we are interested in a fixed number of intervals (three, to be precise)

- the three intervals are all independent

- either an interval contains the true median household income (success) or it does not (failure)

- the probability that any particular interval will contain the true median household income is 0.95

Let X be the number of intervals (out of the three) that contain the true median household income (i.e., the number of successes). X should remind you of a special type of random variable whose probability distribution we have studied previously (Hint: Look at Section 1 of Chapter 5). How do you calculate probabilities for X?

a) We want the probability of three successes. What is this probability?

b) We want the probability of at least two successes. What is this probability?

Exercise 6.27

KEY CONCEPTS - confidence intervals, interpreting confidence intervals

a) The problem tells us that the *estimate* of the population proportion based on the random sample is 54% and that the *margin of error* is 3% for a 95% confidence interval.   Based on the formula,

$$\text{estimate} \pm \text{margin of error}$$

give the 95% confidence interval.

b) Use the confidence interval you computed in part (a) to help you formulate your answer.

COMPLETE SOLUTIONS

Exercise 6.5

a) We are given that $\sigma = \$200$ and $n = 100$. Hence $\sigma_{\bar{x}} = \dfrac{\sigma}{\sqrt{n}} = \dfrac{\$200}{\sqrt{100}} = \$20.$

b) The probability is 0.95 that $\bar{x}$ is within $2 \times \sigma_{\bar{x}}$ of the population mean μ.

c) $2 \times \sigma_{\bar{x}} = \$40$, so about 95% of all samples will capture the true mean of all invoices in the interval $\bar{x}$ plus or minus $40.

Exercise 6.19

a) From the statement of the problem we see

$$C = \text{the level of confidence required} = 0.95$$

hence

$z^* =$ the probability is 0.95 that a standard normal random variable lies between $-z^*$ and z^*

$= 1.96$ (see Table D)

We also see that

$$\sigma = \text{population standard deviation} = 12, \text{ and}$$

$$n = \text{the sample size used} = 100$$

Thus the margin of error m is

$$m = z^* \frac{\sigma}{\sqrt{n}} = 1.96 \frac{12}{\sqrt{100}} = 2.352$$

b) We change n to 10 in (a) and find that the margin of error m is now

$$m = z^* \frac{\sigma}{\sqrt{n}} = 1.96 \frac{12}{\sqrt{10}} = 7.438$$

c) Again we have $z^* = 1.96$ and $\sigma = 12$. We are told we want $m \leq 3$, so we use the maximum possible value we would tolerate, namely $m = 3$, in the formula for n and get

$$n = \left(\frac{z^*\sigma}{m}\right)^2 = \left(\frac{(1.96)(12)}{3}\right)^2 = (7.84)^2 = 61.47$$

Rounding up, we see that the smallest value of n that will accomplish our goal is $n = 62$. Since the budget will allow up to 100 students, this is within the limits of the budget.

Exercise 6.26

Referring to Section 1 of Chapter 5 we see that X has a binomial $B(n=3, p=0.95)$ distribution. We use this distribution to calculate the required probabilities. Table C in your text gives binomial probabilities.

a) Here we want the probability that $X = 3$. Table C only gives binomial probabilities for $p \leq 0.50$. Thus to use Table C we have to rewrite the desired probability in terms of the number of failures (which has $p = 0.05$, a value that is given in Table C) We have (look in the portion of Table C with $n = 3$ and $p = 0.05$)

$$P(X = 3) = P(\text{number of failures} = 0) = 0.8574$$

b) Again, rewriting the desired probability (at least 2 successes) in terms of number of failures we have

$$P(X \geq 2) = P(\text{number of failures} \leq 1) = P(0 \text{ failures}) + P(1 \text{ failure})$$

$$= 0.8574 + 0.1354 = 0.9928$$

Exercise 6.27

a) Based on the information given the 95% confidence interval is

$$54\% \pm 3\%$$

which corresponds to the range 51% to 57%.

b) The 95% tells us the probability that the *method* used to construct the interval will produce an interval containing the true value of the proportion of fans that prefer a play-off as an alternative to the BCS. We can be 95% confident that if we knew all the opinions of adults 18 years and over and computed the proportion favoring a play-off, the interval 51% to 57% would contain this proportion. In interpreting this, one should remember that

• This is a 95% confidence interval, hence we cannot be absolutely certain that the majority of adults 18 years and older favor a play-off. However, there is a good evidence that a majority do, but this may only be a slight majority. because the percentage can be as low as 51%.

• This was a survey of adults 18 years and over. Not all of those sampled are necessarily football fans. This makes it difficult to determine if the survey results are an accurate reflection of the attitudes of football fans.

SECTION 6.2

OVERVIEW

Tests of significance and confidence intervals are the two most widely used types of formal statistical inference. A test of significance is done to assess the evidence against the **null hypothesis** H_0 in favor of an **alternative hypothesis** H_a. Typically the alternative hypothesis is the effect that the researcher is trying to demonstrate, and the null hypothesis is a statement that the effect is not present. The alternative hypothesis can either be **one** or **two-sided**.

Tests are usually carried out by first computing a **test statistic**. The test statistic is used to compute a **P-value** which is the probability of getting a test statistic at least as extreme as the one observed, where the probability is computed when the null hypothesis is true. The P-value provides a measure of how incompatible our data is with the null hypothesis, or how unusual it would be to get data like ours if the null hypothesis were true. Since small P-values indicate data that is unusual or difficult to explain under the null hypothesis, we typically reject the null hypothesis in these cases. In this case, the alternative hypothesis provides a better explanation for our data.

Significance tests of the null hypothesis H_0: $\mu = \mu_0$ with either a one or two-sided alternative are based on the test statistic

$$z = \frac{\bar{x} - \mu_0}{\sigma / \sqrt{n}}$$

The use of this test statistic assumes that we have an SRS from a normal population with known standard deviation σ. When the sample size is large, the assumption of normality is less critical because the sampling distribution of $\bar{x}$

is approximately normal. *P*-values for the test based on z are computed using Table A.

When the *P*-value is below a specified value α, we say the results are statistically significant at level α, or we reject the null hypothesis at level α. Tests can be carried out at a fixed significance level by obtaining the appropriate critical value z^* from the bottom row in Table D.

GUIDED SOLUTIONS

Exercise 6.31

KEY CONCEPTS - null and alternative hypotheses

Typically, H_0 is of the form

$$H_0 : \mu = \text{constant}$$

and H_a is of the form

$$H_a : \mu \neq \text{constant (two-sided alternative)}$$

or

$$H_a : \mu > \text{constant (one-sided alternative)}$$

or

$$H_a : \mu < \text{constant (one-sided alternative)}$$

Remember that in many instances (especially with one-sided alternatives) it is easier to begin with H_a, the effect that we are concerned about, and then to set up H_0 as the statement that the effect is absent.

In each example, think carefully about whether H_a should be one-sided or two-sided.

a) We are interested in determining whether or not the scanner accurately measures a phantom that has known mineral density 1.4. The scanner will be accurate if the mean of the population of all possible measurements on the phantom is 1.4. It will be inaccurate if this mean differs from 1.4. Is H_a one- or two-sided? What is H_a? What is H_0?

b) Why was the old form redesigned? What do you hope to show about the new form? Use this to set up H_a and H_0. Is the alternative one or two-sided?

c) What are you concerned about in this example? Set up H_0 and H_a.

Exercise 6.41

KEY CONCEPTS - interpreting P-values

Write your explanation in the space. Refer to the Section Overview in this Study Guide or Section 6.2 in the text if you need a hint.

Exercise 6.49

KEY CONCEPTS - null and alternative hypotheses, carrying out a significance test about a mean

a) What do the researchers hope to show? This will be the alternative. Write down the null and alternative hypotheses.

b) The first step in carrying out the test is to compute the test statistic $z = \dfrac{\bar{x} - \mu_0}{\sigma / \sqrt{n}}$ which measures how far the sample mean is from the hypothesized value μ_0. To find the numerical value of z, you need to determine μ_0, σ and n from the problem, and then compute $\bar{x}$, the mean DRP score from the data given.

$z =$

Once the value of the test statistic z has been determined, the P-value can be computed. The P-value is the probability that the test statistic takes a value at least as extreme as the one observed. In the space provided, write this down as a probability in terms of Z, the standard normal, and then use Table A to evaluate this probability.

P-value $= P(Z \quad) =$

If you are having trouble doing this directly from the meaning of the P-value, refer to the rules for computing P-values given in Section 6.2 in the text and try to understand the rationale behind them.

Now try to interpret your result in plain language.

Exercise 6.53

KEY CONCEPTS - testing hypotheses at a fixed significance level

The value of the test statistic is given as $z = 2.42$. When carrying out the test at a fixed significance level, you can first compute the P-value and then reject the null hypothesis if the P-value is smaller than the significance level given. However, it is more direct to look up the critical value z^* in Table D and compare the value of the test statistic directly to the critical value.

a) The alternative is $\mu > 1.4$, so the P-value is $P(Z > 2.42)$, which represents the probability of a test statistic at least as extreme as the one observed. Compare the P-value to 0.05. What do you conclude?

Using Table D, we look up the critical value corresponding to the tail probability 0.05 (it does not need to be doubled since the test is one-sided), and find the critical value $z^* = 1.645$. We reject if the computed value of the test statistic exceeds this critical value. What do you conclude? When the significance level is fixed, it is easier to use Table D directly.

b) Follow the procedure in (a), but with significance level 1%.

Exercise 6.61

KEY CONCEPTS - calculating P-values

We need to compare the value $z = 1.12$ to the critical values in Table D. Since the test is two-sided, we double the tail area. Its value is between the critical values $z^* = 1.036$ and $z^* = 1.282$. What two significance levels do these critical values correspond to? Remember that the test is two-sided. What can we say about the P-value?

Now use Table A to determine the P-value by computing the area to the left of -1.37. Is this area the P-value? Why or why not? If not, what do you need to do to find the P-value?

Exercise 6.65

KEY CONCEPTS - relationship between two-sided tests and confidence intervals

a) We are interested in determining whether or not the mean amount of sugar, μ, in the hindguts under these conditions is 7 mg. State the null and alternative hypotheses.

H_0: H_a:

A level α two-sided significance test rejects a hypothesis H_0: $\mu = \mu_0$ when μ_0 falls outside a level $1 - \alpha$ confidence interval for μ. In this exercise, we are told that a 95% confidence interval for μ is 4.2 ± 2.3. Thus α corresponds to 0.05 in this case and we can use the confidence interval to conduct a significance test at the 5% level. What is the value of μ_0 and is it outside the 95% confidence interval? Do we reject at the 5% significance level?

b) What is μ_0 in this case? Can we reject H_0: $\mu = 5$ at the 5% significance level?

COMPLETE SOLUTIONS

Exercise 6.31

a) We are interested in determining whether the mean of the population of all possible measurements on the phantom is equal to 1.4 or not. Thus, the alternative hypothesis is two-sided and of the form H_a: $\mu \neq 1.4$. The null hypothesis is H_0: $\mu = 1.4$.

b) The reason for redesigning the form was to reduce the time to complete it. An effect is present if the mean time μ to complete it is less than for the old form. This corresponds to scores below 0. Hence, an effect is present if μ is less than 0 and so the alternative hypothesis is H_a: $\mu < 0$. The null hypothesis is thus H_0: $\mu = 0$ (but one could also express the null hypothesis as H_0: $\mu \geq 0$.)

c) There is no problem if the mean wattage μ is 60. You will be concerned if μ differs from 60. Thus our hypotheses are H_0: $\mu = 60$ and H_a: $\mu \neq 60$.

Exercise 6.41

In the sample selected by the psychologist, ethnocentrism among church attendees was higher than among nonattenders. Furthermore, the chance of obtaining a difference as large as that observed by the psychologist is less than 0.05 if, in fact, there is no real difference in the population from which the sample was selected. We would take this as strong evidence that ethnocentrism is higher among church attendees than among nonattenders in the population from which the sample was selected.

Exercise 6.49

a) The researchers hope to show that their district mean exceeds the national mean of 32. The alternative is $\mu > 32$, so the hypotheses of interest to the researchers are H_0: $\mu = 32$ and H_a: $\mu > 32$.

b) The value of the test statistic is $z = \dfrac{\bar{x} - \mu_0}{\sigma / \sqrt{n}} = \dfrac{35.091 - 32}{11 / \sqrt{44}} = 1.86$. Since the alternative is $\mu > 32$, the P-value is the chance of getting a value of $\bar{x}$ at least as large as 35.091 if the true mean were 32. In terms of the test statistic, this is equivalent to computing $P(Z \geq z) = P(Z \geq 1.86) = 1 - 0.9686 = 0.0314$.

There is evidence that the mean score of all third graders in this district exceeds the national mean of 32.

Exercise 6.53

a) The value of the test statistic is given as $z = 2.42$. The *P*-value is $P(Z > 2.42)$ = 1- 0.9922 = 0.0078. Since the value is below 0.05, we reject at the 5% level of significance.

Using Table D, we look up the critical value corresponding to the tail probability 0.05 and find the critical value $z^* = 1.645$. A 5% level test rejects if $z > 1.645$. Since 2.42 exceeds this value, we reject at the 5% level of significance.

b) The *P*-value was computed in (a) to be 0.0078. Since 1% = 0.01 exceeds this value, we also reject at the 1% level of significance. This is the advantage of giving the *P*-value. It allows us to assess significance at any level.
Using Table D, we look up the critical value corresponding to the tail probability 0.01 and find the critical value $z^* = 2.326$. Since $z = 2.42$ exceeds the critical value, we reject at the 1% level of significance.

Exercise 6.61

From Table D, $z^* = 1.036$ corresponds to a significance level of 2 x 0.15 = 0.30 since the test is two-sided. The tabled value $z^* = 1.282$ corresponds to a significance level of $2 \times 0.10 = 0.20$. We would reject at $\alpha = .30$ since $|z| = 1.12 > 1.036$, but not at $\alpha = 0.20$ since $|z| = 1.12 < 1.282$. So the *P*-value lies between 0.30 and 0.20.
Using Table A the area to the right of 1.12 is 0.1314 and the *P*-value is $2 \times 0.1314 = 0.2628$ since the test is two-sided. Clearly this lies between the values we found in Table D.

Exercise 6.65

a) The hypotheses are

$$H_0: \mu = 7 \qquad\qquad H_a: \mu \neq 7$$

In this case $\mu_0 = 7$ falls outside the level 95% confidence interval for μ which is $(1.9, 6.5)$. So we reject H_0 at significance level 5%.

c) Since $H_0: \mu = 5$, $\mu_0 = 5$ which doesn't fall outside the 95% confidence interval for μ. So we fail to reject H_0 at significance level 5%.

SECTION 6.3

OVERVIEW

When describing the outcome of a hypothesis test it is more informative to give the P-value than just the reject or not decision at a particular significance level α. The traditional levels of 0.01, 0.05 and 0.10 are arbitrary and serve as rough guidelines.

When testing hypotheses with a very large sample, the P-value can be very small for effects that may not be of practical interest. Don't confuse small P-values with large or important effects. Plot the data to display the effect you are trying to show, and also give a confidence interval which says something about the size of the effect.

Just because a test is not statistically significant doesn't imply that the null hypothesis is true. This may occur when the test is based on a small sample size and has low power. Finally, if you run enough tests, you will invariably find statistical significance for one of them. Be careful in interpreting the results when testing many hypotheses on the same data.

GUIDED SOLUTIONS

Exercise 6.69

KEY CONCEPTS - significance levels vs. P-values

a) In this case, we reject the null hypothesis if the value of the z statistic exceeds 1.645 since the significance level is 0.05. Compute the test statistic when $\bar{x} = 541.4$. What do you conclude at $\alpha = 0.05$?

b) Compute the test statistic when $\bar{x} = 541.5$. What do you conclude at $\alpha = 0.05$?

Exercise 6.73

KEY CONCEPTS - statistical significance vs. practical importance

What would you consider to be a "strong" (practically important) effect? Is the P-value sufficient to assess whether the effect is strong?

Exercise 6.79

KEY CONCEPTS - multiple tests, the *Bonferroni* procedure

If you perform k tests and want protection at level α, use α/k as your cutoff for statistical significance. In our case, $k = 6$ and $\alpha = 0.05$ so each test is conducted at the $0.05/6 = 0.00833$ significance level. Which P-values given lead to rejection at the 0.00833 level?

COMPLETE SOLUTIONS

Exercise 6.69

In this problem we see that the P-value is more informative than just stating whether we accept or reject at $\alpha = 0.05$. In both (a) and (b) the evidence against the null hypothesis is almost the same, yet at the 5% level of significance we reject in (b) but not in (a). Don't attach too much importance to the 5% level of significance. Use the P-value and think about the problem in the context of the field and in terms of other data that may have been collected.

a) $z = \dfrac{\bar{x} - \mu_0}{\sigma/\sqrt{n}} = \dfrac{541.4 - 525}{100/\sqrt{100}} = 1.64$, so we don't reject at $\alpha = 0.05$. The P-value is 0.0505.

b) $z = \dfrac{\bar{x} - \mu_0}{\sigma/\sqrt{n}} = \dfrac{541.5 - 525}{100/\sqrt{100}} = 1.65$, so we reject at $\alpha = 0.05$. The P-value is 0.0495.

Exercise 6.73

Intuitively, we would consider an effect to be "strong" if there were a large difference in the percents of subjects in the two groups who were free of colds. For most people, "large" probably means a difference of more that just a few percent Unfortunately, we are not given the size of the difference, only the P-value. Hence, it is not possible to conclude (from the information given) that vitamin C has a strong effect in preventing colds. All we can conclude is that we are not likely to have observed a difference as large or larger than was actually observed (whatever its magnitude) by chance.

Exercise 6.79

If the P-value is less than $\alpha/k = 0.05/6 = 0.00833$, then we reject using the *Bonferroni* procedure. In this case we reject the null hypothesis for two of the tests - the ones with P-values of 0.008 and 0.001 because these P-values are smaller than 0.00833.

SECTION 6.4

OVERVIEW

The **power** of a significance test is always calculated at a specific alternative hypothesis and is the probability that the test will reject H_0 when that alternative is true. This calculation requires knowledge of the sampling distribution under the specific alternative hypothesis of the test statistic used. Power is usually interpreted as the ability of a test to detect an alternative hypothesis or as the sensitivity of a test to an alternative hypothesis. The power of a test can be increased by increasing the sample size when the significance level remains fixed.

To compute the power of a significance test about a mean of a normal population, we need to:

- state H_0, H_a (the particular alternative we want to detect), and the significance level α,

- find the values of $\bar{x}$ that will lead us to reject H_0,

- calculate the probability of observing these values of $\bar{x}$ when the alternative is true.

Statistical inference can be regarded as giving rules for making decisions in the presence of uncertainty. From this **decision theory** point of view, H_0 and H_a are just two statements of equal status that we must decide between. Decision analysis chooses a rule for deciding between H_0 and H_a on the basis of the probabilities of the two types of errors that we can make. A **Type I error** occurs if H_0 is rejected when it is in fact true. A **Type II error** occurs if H_0 is accepted when in fact H_a is true.

There is a clear relation between α level significance tests and testing from the decision making point of view. The significance level α is the probability of a Type I error and the power of the test against a specific alternative is 1 minus the probability of a Type II error for that alternative.

SAMPLE PROBLEMS

GUIDED SOLUTIONS

Exercise 6.83

KEY CONCEPTS - power of a significance test

To compute the power of a significance test about a mean, we need to:

    (i) state H_0, H_a (the particular alternative we want to detect), and the significance level α,

    (ii) find the values of $\bar{x}$ that will lead us to reject H_0,

    (iii) calculate the probability of observing these values of $\bar{x}$ when the alternative is true.

Following these steps we notice

    (i) In this problem the hypotheses are

$$H_0: \mu = 450 \text{ vs. } H_a: \mu > 450$$

The particular alternative we want to detect is $\mu = 462$. The significance level is $\alpha = 0.01$.

    (ii) The values of $\bar{x}$ that will lead us to reject H_0 are indicated in the problem and are those for which

$$z = \frac{\bar{x} - 450}{100 / \sqrt{500}} \geq 2.326$$

since $\sigma = 100$ and the sample size $n = 500$. Solve the above inequality for one of the form $\bar{x} \geq$ _____.

    (iii) The probability of observing these values of $\bar{x}$ when the alternative is true is

$$P(\bar{x} \geq 460.4 \text{ when } \mu = 462) = P(\frac{\bar{x} - \mu}{\sigma/\sqrt{n}} \geq \frac{460.4 - 462}{100/\sqrt{500}})$$

Now complete the calculation of this probability using Table A. The result will be the desired power.

Exercise 6.87

KEY CONCEPTS - Type I and Type II error probabilities

a) From the information given in the problem we are told that our population is normal, our sample size is $n = 16$, $\sigma = 1$, we are testing

$$H_0: \mu = 0 \text{ vs. } H_a: \mu > 0$$

and we reject H_0 if $\bar{x} > 0$. Thus

$$P(\text{Type I error}) = P(\text{our test rejects } H_0 \text{ when in fact } \mu = 0)$$

$$= P(\bar{x} > 0 \text{ when in fact } \mu = 0)$$

Now compute this probability.

b) In this part we want (since we accept H_0 if $\bar{x} \leq 0$)

$$P(\text{Type II error}) = P(\text{our test accepts } H_0 \text{ when in fact } \mu = 0.2)$$
$$= P(\bar{x} \leq 0 \text{ when in fact } \mu = 0.2)$$

Now compute this probability.

c) See if you can do this on your own. Repeat the argument in (b) with $\mu = 0.6$ in place of $\mu = 0.2$.

COMPLETE SOLUTIONS

Exercise 6.83

We follow the three steps indicated in the Guided Solutions.

(i) This step was completely discussed in the Guided Solutions.

(ii) In terms of $\bar{x}$, after solving the inequality given in the Guided Solutions, we reject H_0 if $\bar{x} \geq (2.326 \times 100 / \sqrt{500}) + 450 = 460.4$.

(iii) We find

$$P(\bar{x} \geq 460.4 \text{ when } \mu = 462) = P(\frac{\bar{x} - \mu}{\sigma / \sqrt{n}} \geq \frac{460.4 - 462}{100 / \sqrt{500}})$$

$$= P(Z \geq -0.36) = 0.6406$$

where we have used Table A to compute $P(Z \geq -0.36)$. The desired power is 0.6406.

Exercise 6.87

a) We have

$$P(\text{Type I error}) = P(\text{our test rejects } H_0 \text{ when in fact } \mu = 0)$$

$$= P(\bar{x} > 0 \text{ when in fact } \mu = 0)$$

$$= P(\frac{\bar{x} - \mu}{\sigma / \sqrt{n}} > \frac{0 - 0}{1 / \sqrt{16}})$$

$$= P(Z > 0)$$

which we know is 0.5.

b) We have

$$P(\text{Type II error}) = P(\text{our test accepts } H_0 \text{ when in fact } \mu = 0.2)$$

$$= P(\bar{x} \leq 0 \text{ when in fact } \mu = 0.2)$$

$$= P(\frac{\bar{x} - \mu}{\sigma / \sqrt{n}} \leq \frac{0 - 0.2}{1 / \sqrt{16}})$$

$$= P(Z \leq -0.8)$$

$$= 0.2119$$

c) We have

$$P(\text{Type II error}) = P(\text{our test accepts } H_0 \text{ when in fact } \mu = 0.6)$$

$$= P(\bar{x} \leq 0 \text{ when in fact } \mu = 0.6)$$

$$= P(\frac{\bar{x} - \mu}{\sigma / \sqrt{n}} \leq \frac{0 - 0.6}{1/\sqrt{16}})$$

$$= P(Z \leq -2.4)$$

$$= 0.0082$$

CHAPTER 7

INFERENCE FOR DISTRIBUTIONS

SECTION 7.1

OVERVIEW

Confidence intervals and significance tests for the mean μ of a normal population are based on the sample mean $\bar{x}$ of an SRS. When the sample size n is large, the central limit theorem suggests that these procedures are approximately correct for other population distributions. In Chapter 6, we considered the (unrealistic) situation in which we knew the population standard deviation σ. In this section, we consider the more realistic case where σ is not known and we must estimate σ from our SRS by the sample standard deviation s. In Chapter 6 we used the **one-sample z statistic**

$$z = \frac{\bar{x} - \mu}{\sigma/\sqrt{n}}$$

which has the $N(0,1)$ distribution. Replacing σ by s, we now use the **one-sample t statistic**

$$t = \frac{\bar{x} - \mu}{s/\sqrt{n}}$$

which has the **t distribution** with n - 1 **degrees of freedom**.

For every positive value of k., there is a t distribution, with k degrees of freedom, denoted $t(k)$. All are symmetric, bell-shaped distributions, similar in shape to normal distributions but with greater spread. As k increases, $t(k)$ approaches the $N(0,1)$ distribution.

A level C **confidence interval for the mean** μ of a normal population when σ is unknown is

$$\bar{x} \pm t^* \frac{s}{\sqrt{n}}$$

where t^* is the upper $(1-C)/2$ critical value of the $t(n-1)$ distribution whose value can be found in Table D of the text or from statistical software. The quantity $t^* \frac{s}{\sqrt{n}}$ is the **margin of error**.

Significance tests of H_0: $\mu = \mu_0$ are based on the one-sample t statistic. Such tests are often referred to as **one-sample t tests**. P-values or fixed significance levels are computed from the $t(n-1)$ distribution using Table D or, more commonly in practice, using statistical software.

The power of the t test is calculated like that of the z test as described in Chapter 6, using an approximate value (perhaps based on past experience or a pilot study) for both σ and s.

One application of these one-sample t procedures is to the analysis of data from **matched pairs** studies. We compute the differences between the two values of a matched pair (often before and after measurements on the same unit) to produce a single sample value. The sample mean and standard deviation of these differences are computed. Depending on whether we are interested in a confidence interval or a test of significance concerning the difference in the population means of matched pairs, we either use the one-sample confidence interval or the one-sample significance test based on the t statistic.

For larger sample sizes, the t procedures are fairly **robust** against nonnormal populations. As a rule of thumb, t procedures are useful for nonnormal data when $n \geq 15$ unless the data show outliers or strong skewness, and for samples of size $n \geq 40$ t procedures can be used for even clearly skewed distributions. Data consisting of small samples from skewed populations can sometimes be analyzed by first applying a **transformation** (such as logarithms) to the data to obtain an approximately normally distributed variable. The t procedures can then be applied to the transformed data. When transformations are used, it is a good idea to examine stemplots, histograms, or normal quantile plots of the transformed data to verify that the transformed data appear to be approximately normally distributed.

Another procedure that can be used with smaller samples from a nonnormal population is the **sign test**. The sign test is most useful for testing for "no treatment effect" in matched pairs studies. As with the matched pairs t test, one computes the differences of the two values in each matched pair. Pairs with difference 0 are ignored and the number of trials n is the count of the remaining pairs. The test statistic is the count X of pairs with a positive difference. *P*-values for *X* are based on the binomial $B(n, 1/2)$ distribution. The sign test is less powerful than the *t* test in cases where the use of the *t* test is justified.

SAMPLE PROBLEMS

GUIDED SOLUTIONS

Exercise 7.9

KEY CONCEPTS - confidence intervals based on the one-sample *t* statistic, checking assumptions, robustness of *t* procedures

a) A stemplot is most easily constructed using statistical software. Either use statistical software or complete the stemplot below. We have used split stems because of the large number of observations between 10 and 19. The skewness of the distribution and the presence of outliers should be readily apparent from the stemplot.

```
1 |
1 |
2 |
2 |
3 |
3 |
4 |
4 |
5 |
5 |
6 |
6 |
7 |
7 |
```

b) To compute a 95% confidence interval we use the formula

$$\bar{x} \pm t^* \frac{s}{\sqrt{n}}$$

where t^* is the upper $(1-C)/2$ critical value of the $t(n-1)$ distribution. Here C is the confidence level of 0.95, and n is the sample size of 50. Thus in this exercise, t^* is the upper $(1-0.95)/2 = 0.025$ critical value of the $t(50-1) = t(49)$ distribution. Find this critical value in Table D.

$t^* =$

Next compute $\bar{x}$ and s, preferably using statistical software or a calculator.

$\bar{x} =$ $s =$

Finally, substitute all these values into the formula to complete the computation of the 95% confidence interval.

$$\bar{x} \pm t^* \frac{s}{\sqrt{n}} =$$

How does your interval compare to the bootstrap intervals and what does this tell you about the robustness of the t interval?

Note that a much easier approach would be to use statistical software to compute the 95% confidence interval directly using the raw data. Consult the manual for your software for instructions.

Exercise 7.11

KEY CONCEPTS - critical values of the $t(k)$ distribution

For confidence intervals, the critical value t^* is the upper $(1-C)/2$ critical value of the $t(n-1)$ distribution. In this problem, $C = 0.95$ and n corresponds to the different sample sizes given. Statistical software will require that you specify the probability associated with the critical value and the degrees of freedom for the t distribution to obtain t^*. Complete the table below and then graph the values on the axes provided. What happens to t^* as n increases? We have provided two critical values for $n = 10$ and $n = 500$.

n	t^*
10	2.2622
20	
30	
40	
50	
100	
200	
500	1.9647

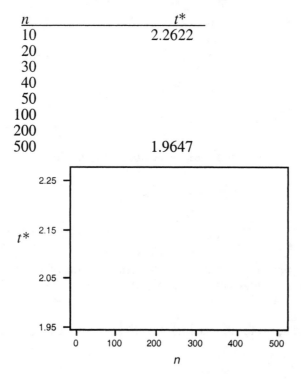

Exercise 7.35

KEY CONCEPTS - one sample t tests

We are interested in whether or not the patient's mean level μ falls above 4.8. What effect do you wish to show? Use this to set up the null and alternative hypotheses

H_0:
H_a:

Make sure you understand why we use a one-sided alternative here.

The sample mean and standard deviation for the data given Exercise 7.34 are easily calculated to be

$$\bar{x} = 5.367 \qquad s = 0.665$$

and the standard error is

$$\frac{s}{\sqrt{n}} = \frac{0.665}{\sqrt{6}} = 0.271.$$

Use this information to calculate the one sample t statistic given below.

$$t = \frac{\bar{x} - \mu}{s/\sqrt{n}} =$$

Based on the value of this t statistic, between which levels from Table D does the P-value lie? What conclusions do you draw?

Exercise 7.40

KEY CONCEPTS - matched pairs experiments, one-sample t tests

a) This is a matched pairs experiment. The matched pair of observations are the right and left thread measurements on each subject. To avoid confounding with time, we would probably want subjects to use both knobs in the same session. We would also want to randomize which knob the subject uses first. How might you do this randomization?

b) The project hopes to show that right-handed people find right-hand threads easier to use than left-hand threads. In terms of the mean μ for the population of differences (left thread time) - (right thread time)

what do we wish to show? This would be the alternative. What are H_0 and H_a (in terms of μ)?

c) For data from a matched pairs study, we compute the differences between the two values of a matched pair to produce a single sample value. These differences are given below for our data.

Right thread	Left thread	Difference = Left - Right
113	137	24
105	105	0
130	133	3
101	108	7
138	115	-23
118	170	52
87	103	16
116	145	29
75	78	3
96	107	11
122	84	-38
103	148	45
116	147	31
107	87	-20
118	166	48
103	146	43
111	123	12
104	135	31
111	112	1
89	93	4
78	76	-2
100	116	16
89	78	-11
85	101	16
88	123	35

The sample mean and standard deviation of these differences are now computed.

$\bar{x} =$ $\qquad$ $s =$

We then use the one sample significance test based on the t statistic. What value of μ should be used

$$t = \frac{\bar{x} - \mu}{s/\sqrt{n}} =$$

From the value of the t statistic and Table D (or using statistical software), the P-value can be computed.

P-value =

What conclusion do you draw?

Note: This problem is most easily done directly using statistical software. The software will compute the differences, the t statistic and the P-value for you. Consult your users manual to see how to do one-sample t tests.

Exercise 7.41

KEY CONCEPTS - matched pairs experiments, confidence intervals

You are required to find a confidence interval for the mean μ of the population of differences

(left thread time) - (right thread time)

The formula for the confidence interval is

$$\bar{x} \pm t^* \frac{s}{\sqrt{n}},$$

where $\bar{x} = 13.32$ and $s = 22.94$ are the mean and standard deviation for the differences obtained in problem 7.40. The critical value t^* is the upper $(1-C)/2$ critical value of the $t(n-1)$ distribution, which can be obtained from Table D or using statistical software. Substitute all these values into the formula to complete the computation of the 95% confidence interval.

$$\bar{x} \pm t^* \frac{s}{\sqrt{n}} =$$

Find the mean time for right-hand threads as a percent of the mean time for left-hand threads. Does the result seem to be of practical importance?

Exercise 7.45

KEY CONCEPTS - appropriateness of statistical procedures

Statistical procedures use information in a sample to make inferences about parameters of a population. Ask yourself, what is the population and the parameter of interest in this exercise? What does our data allow us to say about this parameter?

Exercise 7.48

KEY CONCEPTS - power of the one-sample t test

a) You may wish to refer to Section 6.4 to review power. We wish to determine the power of the test against the alternative $\mu = 0.6$ when the significance level is 0.05 when $n = 12$. We use 0.92 as an estimate of both the population standard deviation σ and s in future samples. The t test with 12 observations rejects H_0: $\mu = 0$ (this is the null hypothesis that the mean difference in yields is 0) if the t statistic

$$t = \frac{\bar{x} - 0}{s/\sqrt{12}}$$

exceeds the upper 5% point of $t(11)$, which is 1.812. This critical value can be obtained using statistical software or using Table D. Taking $s = 0.92$, the event that the test rejects H_0 is

$$t = \frac{\bar{x}}{0.92/\sqrt{12}} \geq 1.812$$

or

$$\bar{x} \geq 1.812 \frac{0.92}{\sqrt{12}} = 0.481$$

The power is the probability that $\bar{x} \geq 0.481$ when $\mu = 0.6$. Taking $\sigma = 0.92$, this probability is found by standardizing $\bar{x}$,

$$P(\bar{x} \geq 0.481 \text{ when } \mu = 0.6) \quad = P(\frac{\bar{x} - 0.6}{0.92/\sqrt{12}} \geq \frac{0.481 - 0.6}{0.92/\sqrt{12}})$$

$$= P(Z \geq -0.45) = 1 - 0.3264 = 0.6736$$

Thus the power is 0.6736.

b) Try this one on your own. Use the same argument as in (a), but now with $n = 30$ rather than $n = 12$ wherever the value of n appears. All other features of the problem are the same.

Exercise 7.49

KEY CONCEPTS - the sign test

a) In Exercise 7.29 we investigated whether the mean μ for the population of differences

(spatial temporal reasoning score after 6 months) - (spatial temporal reasoning score initially)

for matched pairs was greater than 0. Recall that this led to the hypotheses

$$H_0: \mu = 0 \text{ and } H_a: \mu > 0$$

Try to formulate analogous hypotheses in terms of

   • the median for the population of these differences

and in terms of

   • the probability p of an improvement in the test score

What do the statements $H_0: \mu = 0$ and $H_a: \mu > 0$ imply about the median for the population of differences? Remember, the median and mean both attempt to measure the center of a population distribution.

What do the statements $H_0: \mu = 0$ and $H_a: \mu > 0$ imply about the probability p of having an improved test score after 6 months?

b) Recall that for the sign test, pairs with difference 0 are ignored and the number of trials n is the count of the remaining pairs. The test statistic is the count X of pairs with a positive difference, where the differences (changes) are provided in Exercise 7.29. P-values for X are based on the binomial $B(n, 1/2)$ distribution.

How many pairs with difference 0 are there in the data? What, therefore, is n?

What is the observed number of pairs with a positive difference in our data?

In terms of the probability p of completing the task faster with the right-hand thread, we wish to test the hypotheses $H_0: p = 1/2$ and $H_a: p > 1/2$. This implies that the P-value is

   $P(X \geq$ observed number of pairs with a positive difference in our data).

Compute this probability. Note that although X has the binomial $B(n, 1/2)$ distribution, since n is larger than 20, we can use the normal approximation to the binomial distribution to compute the P-value.

COMPLETE SOLUTIONS

Exercise 7.9

a) A stemplot of the data is given below. We have used split stems because of the large number of observations between 10 and 19.

```
1 | 01233344
1 | 5566667778999999
2 | 00124444
2 | 5555566667
3 | 244
3 | 5
4 | 1
4 | 8
5 |
5 |
6 | 3
6 |
7 |
7 | 9
```

b) The desired quantities are

$$t^* = 2.0096 \qquad \bar{x} = 23.56 \qquad s = 12.52$$

$$\bar{x} \pm t^* \frac{s}{\sqrt{n}} = 23.56 \pm (2.0096)\frac{12.52}{\sqrt{50}} = 23.56 \pm 3.56 = (20.00, 27.12)$$

The closeness of this interval to those obtained using the bootstrap illustrates the robustness of the t procedures for $n = 50$, despite the skewness and outlying observations.

Exercise 7.11

The required t^* values are reproduced below along with a graph of t^* versus the sample size. As we can see from the graph, the values of t^* level off as the sample size increases, and in fact they level off to the corresponding z critical value of 1.96.

n	t*
10	2.2622
20	2.0930
30	2.0452
40	2.0227
50	2.0096
100	1.9842
200	1.9720
500	1.9647

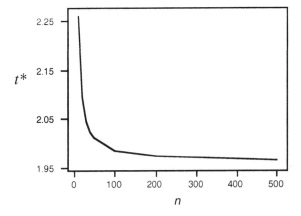

Exercise 7.35

The hypotheses we wish to test are:

$$H_0: \mu = 4.8$$

$$H_a: \mu > 4.8.$$

We compute

$$t = \frac{\bar{x} - \mu}{s/\sqrt{n}} = \frac{5.367 - 4.8}{0.271} = 2.092$$

According to the row corresponding to $n - 1 = 5$ in Table D, the P-value lies between 0.05 and 0.025. Using statistical software we find the P-value is 0.0453. We conclude that there is good evidence that the patient's mean level μ in fact falls above 4.8.

Exercise 7.40

a) The randomization might be carried out by simply flipping a fair coin. If the coin comes up heads, use the right-hand threaded knob first. If the coin comes up tails, use the left-hand threaded knob first.

b) In terms of μ, the mean of the population of differences, (left thread time) - (right thread time), we wish to test if the times for the left threaded knobs are longer than for the right threaded knobs, i.e.

$$H_0: \mu = 0 \text{ vs. } H_a: \mu > 0$$

c) For the 25 differences we compute

$$\bar{x} = 13.32 \qquad s = 22.94$$

We then use the one sample significance test based on the t statistic.

$$t = \frac{\bar{x} - \mu}{s/\sqrt{n}} = \frac{13.32 - 0}{22.94/\sqrt{25}} = 2.903$$

From the value of the t statistic and Table D the P-value is between 0.0025 and 0.005. Using statistical software the P-value is computed as P-value = 0.0039.

We conclude that there is strong evidence that the time for left-hand threads is greater than for right-hand threads on average.

Exercise 7.41

We have $\bar{x} = 13.32$, $s = 22.94$ and $t^* = 1.711$ as the upper 0.95 percentile of the $t(24)$ distribution. This gives

$$\bar{x} \pm t^* \frac{s}{\sqrt{n}} = 13.32 \pm (1.711)\frac{22.94}{\sqrt{25}} = 13.32 \pm 7.85$$

as the 90% confidence interval. The mean for right-hand threads is 104.12 seconds and for the left-hand threads is 117.44. Thus the mean for right-hand threads as a percent of the mean time for left-hand threads is 104.12/117.44 = 88.7%, a substantial savings in time.

Exercise 7.45

The parameter of interest is the mean number of medical doctors per 100,000 people in the population consisting of all 50 states. Since our data is the number of medical doctors per 100,000 people for each of the 50 states, we can average these numbers to compute the desired parameter exactly. No inference is needed.

Exercise 7.48

a) A complete solution is given in the Guided Solutions.

b) The t test with 30 observations rejects H_0: $\mu = 0$ (this is the null hypothesis that the mean difference in yields is 0) if the t statistic

$$t = \frac{\bar{x} - 0}{s/\sqrt{30}}$$

exceeds the upper 5% point of $t(29)$, which is 1.699 (obtained from Table D). Using $s = 0.92$, the event that the test rejects H_0 is

$$t = \frac{\bar{x}}{0.92/\sqrt{30}} \geq 1.699$$

or

$$\bar{x} \geq 1.699 \frac{0.92}{\sqrt{30}} = 0.285$$

The power is the probability that $\bar{x} \geq 0.285$ when $\mu = 0.6$. Taking $\sigma = 0.92$, this probability is found by standardizing $\bar{x}$,

$$P(\bar{x} \geq 0.285 \text{ when } \mu = 0.6) \quad = P(\frac{\bar{x} - 0.6}{0.92/\sqrt{30}} \geq \frac{0.285 - 0.6}{0.92/\sqrt{30}})$$

$$= P(Z \geq -1.88) = 1 - 0.0301 = 0.9699.$$

Thus the power is 0.9699.

Exercise 7.49

a) In terms of the median for the population of differences

(spatial temporal reasoning score after 6 months) - (spatial temporal reasoning score initially)

the hypotheses are

H_0: population median = 0 and H_a: population median > 0

since these hypotheses test whether there is an improvement in test score.

In terms of the probability p of an improvement in test score, the hypotheses would be

$$H_0: p = 1/2 \text{ and } H_a: p > 1/2$$

b) There are 2 pairs with difference 0 in the data, thus $n = 32$. The number of pairs with a positive difference in our data is 28. Thus the P-value is

$$P(X \geq 28) = 1 - P(X \leq 27)$$

where X has the binomial $B(32, 1/2)$ distribution. We use the normal approximation to the binomial to compute this probability. X has mean

$$\mu = np = (32)(1/2) = 16$$

and standard deviation

$$\sigma = \sqrt{np(1-p)} = \sqrt{(32)(1/2)(1/2)} = \sqrt{8} = 2.83.$$

The z-score of 27 is thus

$$z = \frac{27 - \mu}{\sigma} = \frac{27 - 16}{2.83} = 3.89$$

and so our P-value is

$$
\begin{aligned}
P(X \geq 28) \quad &= 1 - P(X \leq 27) \\
&= 1 - P(\text{standard normal random variable} \leq 3.89) \\
&< 0.0002
\end{aligned}
$$

The probability using the binomial distribution and statistical software is, to four decimals, 0.0000.

SECTION 7.2

OVERVIEW

One of the most commonly used significance tests is the **comparison of two population means,** μ_1 and μ_2. In this setting we have two distinct, independent SRS from two populations or two treatments on two samples. The procedures are based on the difference $\bar{x}_1 - \bar{x}_2$. When the populations are not normal, the results obtained using the methods of this section are approximately correct due to the central limit theorem.

Tests and confidence intervals for the difference in the population means, $\mu_1 - \mu_2$, are based on the **two-sample t statistic,**

$$t = \frac{(\bar{x}_1 - \bar{x}_2) - (\mu_1 - \mu_2)}{\sqrt{\dfrac{s_1^2}{n_1} + \dfrac{s_2^2}{n_2}}}$$

Despite the name, this test statistic does *not* have an exact t distribution. However there are good approximations to its distribution which allow us to carry out valid significance tests. Conservative procedures use the $t(k)$

distribution as an approximation where the degrees of freedom k is taken to be the smaller of n_1 -1 and n_2 -1. More accurate procedures use the data to estimate the degrees of freedom k. This is the procedure that is followed by most statistical software.

To carry out a significance test for $H_0:\mu_1 = \mu_2$, use the two-sample t statistic

$$t = \frac{(\bar{x}_1 - \bar{x}_2)}{\sqrt{\dfrac{s_1^2}{n_1} + \dfrac{s_2^2}{n_2}}}$$

The P-value is found by using the approximate distribution $t(k)$, where k is estimated from the data when using statistical software, or can be taken to be the smaller of n_1 -1 and n_2 -1 for a conservative procedure.

An approximate confidence C level **confidence interval** for $\mu_1 - \mu_2$ is given by

$$(\bar{x}_1 - \bar{x}_2) \pm t^* \sqrt{\frac{s_1^2}{n_1} + \frac{s_2^2}{n_2}}$$

where t^* is the upper (1 - C)/2 critical value for $t(k)$, where k is estimated from the data when using statistical software, or can be taken to be the smaller of n_1 -1 and n_2 -1 for a conservative procedure. The procedures are most robust to failures in the assumptions when the sample sizes are equal.

The **pooled two-sample t procedures** are used when we can safely assume that the two populations have equal variances. The modifications in the procedure are the use of the pooled estimator of the common unknown variance

$$s_p^2 = \frac{(n_1 - 1)s_1^2 + (n_2 - 1)s_2^2}{n_1 + n_2 - 2}$$

and critical values obtained from the t $(n_1 + n_2$ -2) distribution.

GUIDED SOLUTIONS

Exercise 7.69

KEY CONCEPTS - tests using the two-sample t, generalizations beyond the data

a) When means and standard deviations for the two samples are given, it is relatively easy to compute the value of the two-sample t statistic. With raw data as in this exercise, the details of the computations are best left to statistical software. However, although you may let the computer do the tedious computations, it is still important to know what all the numbers in the output mean and how to interpret the results. The output below is reproduced from MINITAB. The output is fairly standard between the different software packages.

Two Sample T-Test and Confidence Interval

```
Twosample T for Low vs High
        N      Mean     StDev    SE Mean
Low    14     4.640     0.690      0.18
High   14     6.429     0.430      0.12

T-Test mu Low = mu High (vs not =): T= -8.23 P=0.0000 DF = 21.77
```

The output begins with summary information for the two samples. For the Low group, $n_1 = 14$, $\bar{x}_1 = 4.640$, $s_1 = 0.690$ and the SE Mean is $s_1 / \sqrt{n_1} = 0.18$. For the High group, $n_2 = 14$, $\bar{x}_2 = 6.429$ $s_2 = 0.430$ and the SE Mean is $s_2 / \sqrt{n_2} = 0.12$. The next line gives the hypotheses, test statistic and P-value:

`mu Low = mu High` corresponds to H_0: $\mu_{\text{Low}} = \mu_{\text{High}}$, and

`(vs not =)` corresponds to H_a: $\mu_{\text{Low}} \neq \mu_{\text{High}}$.

Next is $t = \dfrac{\bar{x}_1 - \bar{x}_2}{\sqrt{\dfrac{s_1^2}{n_1} + \dfrac{s_2^2}{n_2}}} = \dfrac{4.640 - 6.429}{\sqrt{\dfrac{(0.690)^2}{14} + \dfrac{(0.430)^2}{14}}} = -8.23$ and the P-value =

0.0000. The degrees of freedom is found using the formula

$$\frac{\left(\dfrac{s_1^2}{n_1} + \dfrac{s_2^2}{n_2}\right)^2}{\dfrac{1}{n_1 - 1}\left(\dfrac{s_1^2}{n_1}\right)^2 + \dfrac{1}{n_2 - 1}\left(\dfrac{s_2^2}{n_2}\right)^2} = \frac{\left(\dfrac{0.690^2}{14} + \dfrac{0.430^2}{14}\right)^2}{\dfrac{1}{14 - 1}\left(\dfrac{0.690^2}{14}\right)^2 + \dfrac{1}{14 - 1}\left(\dfrac{0.430^2}{14}\right)^2} = 21.77$$

You should make sure you understand how all the numbers in the computer output are obtained. Now use the information in the output to answer the questions: Is the difference in mean ego strength significant at the 5% level? At the 1% level? Why?

b) Are these groups random samples from low-fitness and high-fitness groups of middle-aged men? Where did the individuals in the samples come from? How might this affect generalizing to the population of all middle-aged men?

Exercise 7.79

KEY CONCEPTS - tests and confidence intervals using the two-sample t, relationship between two-sided tests and confidence intervals

The summary statistics for the comparison of children and adult VOT scores are reproduced below.

Group	n	$\bar{x}$	s
Children	10	-3.67	33.89
Adults	20	-23.17	50.74

a) Recall that the standard error of the mean is $s/\sqrt{n}$, where s is the sample standard deviation. Use this to compute the standard error of the sample mean VOT for the adults.

The standard error for the difference in two means $\bar{x}_1 - \bar{x}_2$ is $\sqrt{\dfrac{s_1^2}{n_1} + \dfrac{s_2^2}{n_2}}$. Apply this to compute the standard error for the difference between the mean VOT for children and adults. Don't forget to square both standard deviations when evaluating the formula for the standard error.

b) The researchers were interested in a difference in either direction. Use this to set up the null and alternative hypotheses. Find the numerical value of the two-sample t statistic using the formula below. Remember that you found the standard error of the difference between the means in (a) which corresponds to the denominator of the t.

$$t = \frac{\bar{x}_1 - \bar{x}_2}{\sqrt{\dfrac{s_1^2}{n_1} + \dfrac{s_2^2}{n_2}}} =$$

Compute the P-value using a t distribution with the smaller of $n_1 - 1 = 9$ and $n_2 - 1 = 19$ degrees of freedom, and comparing the computed value of t to the critical values given in Table D. Remember to double the tail probabilities since the test is two-sided. State your conclusions.

c) The formula for the 95% confidence interval is $\bar{x}_1 - \bar{x}_2 \pm t^* \sqrt{\dfrac{s_1^2}{n_1} + \dfrac{s_2^2}{n_2}}$, where t^* is the upper $(1-C)/2 = 0.025$ critical value for the t distribution with degrees of freedom equal to the smaller of $n_1 - 1 = 9$ and $n_2 - 1 = 19$. First find t^* from Table D, and complete the computations for the confidence interval. Remember the relationship between two-sided tests and confidence intervals. We reject the null hypothesis $H_0 : \mu_1 - \mu_2 = 0$ at significance level α when the $1 - \alpha$ confidence interval for $\mu_1 - \mu_2$ doesn't contain the value 0.

Exercise 7.86

KEY CONCEPTS - pooled t test

The summary statistics for the comparison of male and female scores on the Chapin Social Insight Test are reproduced below.

Group	Sex	n	$\bar{x}$	s
1	M	133	25.34	5.05
2	F	162	24.94	5.44

The hypotheses are $H_0 : \mu_1 = \mu_2$ and $H_a : \mu_1 \neq \mu_2$. The computations for the pooled two-sample t test require the pooled estimator of σ^2 which combines the estimates of σ^2 from the two samples. The formula is

$$s_p^2 = \frac{(n_1 - 1)s_1^2 + (n_2 - 1)s_2^2}{n_1 + n_2 - 2} =$$

where you need to remember to square the standard deviations. The next step is to compute the value of the test statistic. The formula is

$$t = \frac{\bar{x}_1 - \bar{x}_2}{s_p \sqrt{\dfrac{1}{n_1} + \dfrac{1}{n_2}}} =$$

where you need to remember to use s_p, the square root of s_p^2 in the denominator. The P-value is found by comparing the computed value of t to critical values for the $t(n_1 + n_2 - 2) = t(293)$ distribution, and then doubling the tail area since the alternative is two-sided. Use Table D to arrive at your answer.

COMPLETE SOLUTIONS

Exercise 7.69

a) If we know the P-value, we can assess the significance at any level by comparing the P-value to the significance level. Since the P-value is the smallest α level at which we can reject H_0, if the P-value is smaller than the significance level given we reject H_0; otherwise we do not reject. Since the P-value in the output is given as 0.0000, it is smaller than both 5% and 1%. So we reject H_0 at both these significance levels.

b) The individuals in the study are not random samples from low-fitness and high-fitness groups of middle-aged men. There are two ways in which they might be systematically different. The first is that all subjects in the study are college faculty. The "ego strength" personality factor for low-fitness and high-fitness middle-aged college faculty members might differ from the general population, and so might the mean difference between the groups. In addition, we don't have random samples of college faculty members. We are using volunteers in the study, and they might differ from both college faculty and general population. It is hard to say in which direction this might bias the results, but the possibility of bias is definitely present.

Exercise 7.79

Group	n	$\bar{x}$	s
Children	10	-3.67	33.89
Adults	20	-23.17	50.74

a) The standard error of the sample mean VOT for the adults is

$$s/\sqrt{n} = 50.74/\sqrt{20} = 11.346$$

The standard error for the difference $\bar{x}_1 - \bar{x}_2$ between the mean VOT for children and adults is

$$\sqrt{\frac{s_1^2}{n_1} + \frac{s_2^2}{n_2}} = \sqrt{\frac{33.89^2}{10} + \frac{50.74^2}{20}} = 15.607$$

b) The hypotheses are $H_0:\mu_1 = \mu_2$ versus $H_a:\mu_1 \neq \mu_2$, since the researchers were interested in a difference in either direction. The numerical value of the two-sample t statistic is given below where we have used the fact that the denominator of the t was computed in (a).

$$t = \frac{\bar{x}_1 - \bar{x}_2}{\sqrt{\frac{s_1^2}{n_1} + \frac{s_2^2}{n_2}}} = \frac{-3.67 - (-23.17)}{15.607} = 1.25$$

The smaller of $n_1 - 1 = 9$ and $n_2 - 1 = 19$ is 9 so we can refer the computed value of t to a $t(9)$ in Table D. The value 1.25 is between the critical values corresponding to upper tail probabilities of 0.10 and 0.15. This gives a P-value between 0.20 and 0.30 We need to double upper tail probabilities from the table because the test is two-sided (The exact P-value for a t with 9 degrees of freedom using statistical software is 0.2428) The data give no evidence of difference in mean VOT between children and adults.

c) The 95% confidence interval is $\bar{x}_1 - \bar{x}_2 \pm t^* \sqrt{\frac{s_1^2}{n_1} + \frac{s_2^2}{n_2}}$, where t^* is the upper $(1-C)/2 = 0.025$ critical value for the t distribution with degrees of freedom equal to the smaller of $n_1 - 1 = 9$ and $n_2 - 1 = 19$. Using 9 degrees of freedom we find the value of $t^* = 2.262$, and the confidence interval is

$$-3.67 - (-23.17) \pm 2.262(15.607) = (-15.8, 54.8)$$

Since the interval includes the value 0, we fail to reject H_0 at the 5% level of significance. We knew this from (a) since the P-value exceeded 5%.

Exercise 7.86

The pooled estimator of σ^2 is

$$s_p^2 = \frac{(n_1 - 1)s_1^2 + (n_2 - 1)s_2^2}{n_1 + n_2 - 2} = \frac{(133 - 1)(5.05)^2 + (162 - 1)(5.44)^2}{133 + 162 - 2} = 27.751$$

The value of the pooled two-sample t statistic is

$$t = \frac{\bar{x}_1 - \bar{x}_2}{s_p \sqrt{\frac{1}{n_1} + \frac{1}{n_2}}} = \frac{25.34 - 24.94}{5.268 \sqrt{\frac{1}{133} + \frac{1}{162}}} = 0.649$$

which agrees quite closely with the answer in Example 7.16. When the two standard deviations are close, the numerical values of the two-sample t statistic and the pooled two-sample t statistic are quite similar, as are the conclusions.

The P-value is found by comparing the computed value of t to critical values for the $t(n_1 + n_2 - 2) = t(293)$ distribution, and then doubling the upper tail probability because the alternative is two-sided. Table D shows the value of

0.649 does not reach the critical value corresponding to the upper 0.25 tail probability, which is the largest upper tail probability in the table Doubling this tail area we can see that the P-value exceeds .5. Statistical software can be used to compute the P-value giving the exact result of 0.5168. The data give no evidence of a difference in mean scores between men and women.

SECTION 7.3

OVERVIEW

There are formal inference procedures to compare the standard deviations of two normal populations as well as the two means. The validity of the procedures is seriously affected by nonnormality, and they are not recommended for regular use. The procedures are based on the *F* **statistic,** which is the ratio of the two sample variances

$$F = \frac{s_1^2}{s_2^2}$$

If the data consist of independent simple random samples of sizes n_1 and n_2 from two normal populations, then the F statistic has the F distribution, $F(n_1 - 1, n_2 - 1)$, if the two population standard deviations σ_1 and σ_2 are equal.

The power of the pooled two-sample t test can be found using the noncentral t-distribution or a normal approximation to it. The critical value for the significance test, the degrees of freedom and the noncentrality parameter for alternatives of interest are all required for the computation. Power calculations can be useful when comparing alternative designs and assumptions.

GUIDED SOLUTIONS

Exercise 7.95

KEY CONCEPTS - the F test for equality of the standard deviations of two normal populations

The data are from Exercise 7.71. The data consist of wheat prices in July and September using survey data. In Exercise 7.71, you were asked to convert the standard errors provided into standard deviations. We have reproduced this information in the table below.

Month	n	$s / \sqrt{n}$	s
July	90	$0.023	$0.2182
September	45	$0.029	$0.1945

We are interested in testing the hypotheses $H_0:\sigma_1 = \sigma_2$ and $H_a:\sigma_1 \neq \sigma_2$. The two-sided test statistic is the larger variance divided by the smaller variance. To perform the test, first find the two sample variances and take the ratio of the larger to the smaller.

$$F = \frac{\text{larger } s^2}{\text{smaller } s^2} = $$

If σ_1 and σ_2 are equal, this ratio has the $F(n_1 - 1, n_2 - 1) = F(90-1, 45-1) = F(89, 44)$ distribution. From Table E, what value would the ratio of variances have to exceed to reject the null hypothesis at the 5% level of significance? What do you need to do with the significance level when going to Table E? (Remember it is a two-sided alternative.) What can you say about the P-value using the information in Table E?

Exercise 7.99

KEY CONCEPTS - computing the power of the pooled two-sample t test

Power computations are important when planning a study. You want to ensure you have enough data to detect alternative hypotheses that are of interest to you. Remember, just because a test is not statistically significant doesn't imply that the null hypothesis is true. This situation may occur when the test is based on a small sample size and has low power, or ability, to detect alternatives of interest. Power calculations done before the study help to avoid this problem.

We follow the sequence of steps given in the text, to keep our calculations organized.

1. Specify:

a) An alternative that you consider important. Read through the exercise. What alternative (value of $\mu_1 - \mu_2$) is the researcher interested in detecting?

b) What sample sizes are the researchers planning on using?

c) The type I error.

d) A guess at the standard deviation σ. This is based on previous data. What value are the researchers going to use?

2. Find the df $= n_1 + n_2 - 2$ and the value t^* leading to rejection.

3. Calculate the noncentrality parameter

$$\delta = \frac{|\mu_1 - \mu_2|}{\sigma\sqrt{(1/n_1) + (1/n_2)}} =$$

4. The power is the probability that a noncentral t random variable with 158 degrees of freedom and noncentrality parameter $\delta = 2.92$ will be less than $t^* = 1.65$. If you do not have software to perform this calculation you can approximate the power as $P(Z > t^* - \delta)$.

$P(Z > t^* - \delta) =$

COMPLETE SOLUTIONS

Exercise 7.83

For the data collected in July the sample standard deviation is $0.2182 and the sample variance is $(0.2182)^2 = 0.0476$. For the data collected in September the sample standard deviation is $0.1945, and the sample variance is $(0.1945)^2 = 0.0378$. Computing the test statistic gives

$$F = \frac{\text{larger } s^2}{\text{smaller } s^2} = \frac{0.0476}{0.0378} = 1.26$$

If σ_1 and σ_2 are equal, this ratio has the $F(n_1 - 1, n_2 - 1) = F(90-1, 45-1) = F(89, 44)$ distribution. From Table E, using $F(120, 50)$ we see the upper 0.10 critical value is 1.38, so even at the $2 \times 0.10 = 20\%$ level of significance we would not reject the null hypothesis. Using statistical software, the exact upper tail probability (probability of exceeding 1.26 for an $F(89, 44)$ distribution is 0.2000 and the P-value $= 2 \times 0.2000 = 0.4000$, so there is no evidence of a difference in variability in the two months.

Exercise 7.89

a) The researchers think the true difference in mean birth weights might be about 300 grams and a difference this large is clinically important. The value $\mu_1 - \mu_2 = 300$ is an alternative considered important to detect.

b) The sample sizes are $n_1 = 80$ and $n_2 = 80$.

c) The type I error is $\alpha = 0.05$.

d)

1. Our guess for σ is going to be 650 grams.

2. The df $= n_1 + n_2 - 2 = 158$ and the value $t^* = 1.65$.

3. The noncentrality parameter is

$$\delta = \frac{|\mu_1 - \mu_2|}{\sigma\sqrt{(1/n_1) + (1/n_2)}} = \frac{300}{650\sqrt{(1/80) + (1/80)}} = 2.92$$

4. You can approximate the power as
$$P(Z > t^* - \delta) = P(Z > 1.65 - 2.92) = P(Z > -1.27) = 1 - .1020 = .8980$$

Using SAS the probability can be computed exactly using the noncentral t distribution. The SAS function which does this is called PROBT and to use it you need to specify t^*, DF, and δ. The SAS "command" is

$$POWER = 1 - PROBT(t^*, DF, \delta) = 1 - PROBT(1.65, 158, 2.92) = 0.89745$$

The normal approximation and the exact calculation agree well in this case. Provided the degrees of freedom is not too small (at least 30 - 40), the agreement will be good and the normal approximation will be sufficient for applied work.

With 80 observations in each group, there is a high probability (almost 90%) that our test will be able to detect a difference in means as large as 300 grams. Said another way, if the difference in means in as large as 300 grams, the probability we will reject the null hypothesis, which would be the correct decision, is about 90%. The sample sizes are sufficiently large to detect a meaningful difference.

CHAPTER 8

INFERENCE FOR PROPORTIONS

SECTION 8.1

OVERVIEW

In this section we consider inference about a population proportion p from an SRS of size n based on the **sample proportion** $\hat{p} = X/n$ and the **Wilson estimate** $\tilde{p} = (X + 2)/(n + 4)$, which is the sample proportion with 2 successes and 2 failures added to the data. In these formulas, X is the number of "successes" (occurrences of the event of interest) in the sample. If the population is at least ten times as large as the sample, the individual observations will be approximately independent and X will have a distribution which is approximately binomial $B(n, p)$. While it is possible to develop procedures for inference about p based on the binomial $B(n, p)$ distribution, these can be awkward to work with because of the discrete nature of the binomial distribution. When n is large, we can treat $\hat{p}$ as having a distribution which is approximately normal with mean $\mu = p$ and standard deviation $\sigma = \sqrt{p(1-p)/n}$, and $\tilde{p}$ as having a distribution which is approximately normal with mean $\mu = p$ and standard deviation $\sigma = \sqrt{p(1-p)/(n+4)}$.

An **approximate level C confidence interval** for p is

$$\tilde{p} \pm z^{*}\mathrm{SE}_{\tilde{p}}$$

where z^{*} is the upper $(1-C)/2$ critical value of the standard normal distribution, and

$$SE_{\tilde{p}} = \sqrt{\frac{\tilde{p}(1-\tilde{p})}{n+4}}$$

is the **standard error** of $\tilde{p}$, and $z^* SE_{\tilde{p}}$ is the **margin of error.**

Tests of the hypothesis H_0: $p = p_0$ are based on the z **statistic**

$$z = \frac{\hat{p} - p_0}{\sqrt{\dfrac{p_0(1-p_0)}{n}}}$$

with P-values calculated from the $N(0, 1)$ distribution.

The **sample size** n required to obtain a confidence interval of approximate margin of error m for a proportion is

$$n + 4 = \left(\frac{z^*}{m}\right)^2 p^*(1\text{-}p^*)$$

where p^* is a guessed value for the population proportion and z^* is the upper $(1\text{-}C)/2$ critical value of the standard normal distribution. To guarantee that the margin of error of the confidence interval is less than or equal to m no matter what the value of the population proportion may be, use a guessed value of $p^* = 1/2$, which yields

$$n + 4 = \left(\frac{z^*}{2m}\right)^2$$

SAMPLE PROBLEMS

GUIDED SOLUTIONS

Exercise 8.9

KEY CONCEPTS - confidence intervals for a proportion

Recall that an approximate level C confidence interval for p, the true proportion of the month's orders that were shipped on time is

$$\tilde{p} \pm z^* \sqrt{\frac{\tilde{p}(1-\tilde{p})}{n+4}}$$

where z^* is the upper $(1-C)/2$ critical value of the standard normal distribution. From the information given in the problem, provide the values requested below.

n = sample size =

$\tilde{p}$ = the Wilson estimate of the true proportion of the month's orders shipped on time =

C = level of confidence requested =

z^* = upper $(1-C)/2$ critical value of the standard normal distribution =

You will need to use Table D to find z^*. Now substitute these values into the formula for the confidence interval to complete the problem.

$$\tilde{p} \pm z^* \sqrt{\frac{\tilde{p}(1 - \tilde{p})}{n + 4}} =$$

Exercise 8.13

KEY CONCEPTS - when to use the normal approximation to the binomial test

Recall that the rule of thumb is that the normal approximation to the binomial is appropriate if <u>both</u> $np_0 \geq 10$ and $n(1 - p_0) \geq 10$. These are the conditions that we must check in each of (a) through (d).

Exercise 8.15

KEY CONCEPTS - testing hypotheses about a proportion

Before proceeding, we review some facts. We know that the sampling distribution of any statistic is the distribution of the values of the statistic that we obtain from many, many independent samples. The survey method devised by the market researchers yields a statistic, namely the proportion of Indiana households in urban areas that we obtain using the method. This statistic will have a sampling distribution. Let p denote the mean of this sampling distribution. This is our population parameter in this problem and the proportion $\hat{p} = 0.62$ of urban households obtained in the sample actually conducted is an estimate of p.

We are interested in determining if the method used by the market researchers to collect the survey data gives samples that are representative of the state in regard to rural versus urban residence. In the terminology of statistics, this question is equivalent to asking if the survey method devised by the market researchers yields an unbiased estimate of the "true" proportion of Indiana households that are in urban areas, which is 0.64 based on the census data. In other words, is $p = 0.64$?

a) In terms of a formal test of hypotheses, we want to test

$$H_0:$$
$$H_a:$$

b) Recall that tests of the hypothesis $H_0: p = p_0$ are based on the z statistic

$$z = \frac{\hat{p} - p_0}{\sqrt{\dfrac{p_0(1 - p_0)}{n}}}$$

We have already mentioned that $\hat{p} = 0.62$ here. Identify n and p_0 in this example and calculate z.

$$z = \frac{\hat{p} - p_0}{\sqrt{\dfrac{p_0(1 - p_0)}{n}}} =$$

For our hypotheses, the P-value is

$$2P(\text{standard normal random variable} \geq |z|)$$

(Why is the probability multiplied by 2?)

Calculate this P-value and then summarize your conclusions.

$$P\text{-value} = 2P(\text{standard normal random variable} \geq |z|)$$

$$=$$

Conclusions:

c) The results should be the same as in the previous problem. In fact the z statistics are identical but with the opposite sign. Can you see why that is the case? What is the general conclusion?

Exercise 8.25

KEY CONCEPTS - sample size and margin of error

The sample size n required to obtain a confidence interval of approximate margin of error m for a proportion is

$$n + 4 = \left(\frac{z^*}{m}\right)^2 p^*(1 - p^*)$$

where p^* is a guessed value for the population proportion and z^* is the critical value of the standard normal distribution for the desired level of confidence. To apply this formula here we must determine

m = desired margin of error =

p^* = a guessed value for the population proportion =

C = desired level of confidence =

z^* = the upper $(1-C)/2$ critical value of the standard normal distribution

=

From the statement of the exercise, what are these values? Once you have determined them, use the formula to compute the required sample size n.

$$n + 4 = \left(\frac{z^*}{m}\right)^2 p^*(1 - p^*) =$$

or $n = $ _____.

COMPLETE SOLUTIONS

Exercise 8.9

The desired information is

n = sample size = 200

$\tilde{p}$ = the Wilson estimate of the true proportion of the month's orders shipped on time = $(185 + 2)/(200 + 4) = 0.917$

C = level of confidence requested = 0.95

z^* = upper $(1-C)/2$ critical value of the standard normal distribution
= upper 0.025 critical value of the standard normal distribution
= 1.96.

Substituting these values into the formula for the confidence interval yields

$$\tilde{p} \pm z^* \sqrt{\frac{\tilde{p}(1-\tilde{p})}{n+4}} = 0.917 \pm (1.96)\sqrt{\frac{(0.917)(1-0.917)}{204}}$$

$$= 0.917 \pm (1.96)(0.0193) = 0.917 \pm 0.038$$

$$\text{or } 0.879 \text{ to } 0.955.$$

Exercise 8.13

a) We see that $np_0 = (10)(0.6) = 6 < 10$, so the normal approximation to the binomial should <u>not</u> be used in this case.

b) We see that $np_0 = (100)(0.4) = 40 \geq 10$ and $n(1-p_0) = (100)(1 - 0.4) = (100)(0.6) = 60 \geq 10$. Thus the normal approximation to the binomial can be used in this case.

c) We see that $np_0 = (2000)(0.996) = 1992 \geq 10$, but $n(1-p_0) = (2000)(1 - 0.996) = (2000)(0.004) = 8 < 10$. Thus the normal approximation to the binomial should <u>not</u> be used in this case.

d) We see that $np_0 = (500)(0.25) = 125 \geq 10$ and $n(1-p_0) = (500)(1 - 0.25) = (500)(0.75) = 375 \geq 10$. Thus the normal approximation to the binomial can be used in this case.

Exercise 8.15

a) We want to test

$$H_0: p = 0.64$$
$$H_a: p \neq 0.64$$

b) Here $n = 500$ and $p_0 = 0.62$, so

$$z = \frac{\hat{p} - p_0}{\sqrt{\frac{p_0(1-p_0)}{n}}} = \frac{0.62 - 0.64}{\sqrt{\frac{0.64(1-0.64)}{500}}} = \frac{-0.02}{0.0215} = -0.93.$$

Thus

$$P\text{-value} = 2P(\text{standard normal random variable} \geq |z|)$$
$$= 2P(\text{standard normal random variable} \geq 0.93)$$
$$= 2(1 - P(\text{standard normal random variable} < 0.93))$$

$$= 2(1 - 0.8238)$$

$$= 0.3524$$

Conclusions: The P-value is not very small, suggesting that the data do not provide convincing evidence against H_0. In other words, the survey method used by the market researchers appears to represent the state in regard to rural versus urban residence.

c) Since there are only two outcomes, rural vs. urban, whatever the level of evidence in favor of the rural proportion being correct must be the same as for urban (the general principle is that it doesn't matter which of the two categories is called success - the conclusions agree). In terms of the z statistic, you should notice that the value of the denominator is identical to the denominator used in Exercise 8.14, while the numerator is of the opposite sign.

Exercise 8.25

From the statement of the exercise we have

m = desired margin of error = 0.05

p^* = a guessed value for the population proportion = 0.25

C = desired level of confidence = 0.95

z^* = the upper $(1-C)/2$ critical value of the standard normal distribution

$= 1.96$

The required sample size n is thus

$$n + 4 = \left(\frac{z^*}{m}\right)^2 p^*(1-p^*) = \left(\frac{1.96}{0.05}\right)^2 (0.25)(1 - 0.25) = 288.12$$

which we round up to 289. Thus the required $n = 289 - 4 = 285$.

SECTION 8.2

OVERVIEW

Confidence intervals and tests designed to compare two population proportions are based on the **difference in the sample proportions** $D = \hat{p}_1 - \hat{p}_2$ and the difference in **Wilson estimates** $\tilde{D} = \tilde{p}_1 - \tilde{p}_2$ where

$$\tilde{p}_1 = \frac{X_1 + 1}{n_1 + 2} \quad \text{and} \quad \tilde{p}_2 = \frac{X_2 + 1}{n_2 + 2}$$

and X_1 and X_2 are the number of successes in each group.

The formula for the level C confidence interval is

$$(\tilde{p}_1 - \tilde{p}_2) \pm z^* \text{SE}_{\tilde{D}}$$

where z^* is the upper $(1 - C)/2$ standard normal critical value and $\text{SE}_{\tilde{D}}$ is the standard error for the difference in the two proportions computed as

$$\text{SE}_{\tilde{D}} = \sqrt{\frac{\tilde{p}_1(1 - \tilde{p}_1)}{n_1 + 2} + \frac{\tilde{p}_2(1 - \tilde{p}_2)}{n_2 + 2}}$$

Significance tests for the equality of the two proportions, $H_0: p_1 = p_2$, use a different standard error for the difference in the sample proportions which is based on a **pooled estimate** of the common (under H_0) value of p_1 and p_2,

$$\hat{p} = \frac{X_1 + X_2}{n_1 + n_2}.$$

The test uses the z *statistic*

$$z = \frac{\hat{p}_1 - \hat{p}_2}{\text{SE}_{Dp}}$$

where

$$\text{SE}_{Dp} = \sqrt{\hat{p}(1 - \hat{p})\left(\frac{1}{n_1} + \frac{1}{n_2}\right)}$$

and P-values are computed using Table A of the standard normal distribution.

GUIDED SOLUTIONS

Exercise 8.37

KEY CONCEPTS - confidence intervals for the difference between two population proportions

The data which give the number of job applicants who lied about having a degree over two six month periods are reproduced below with the sample sizes for each group.

Period	n	X(lied)
1	84	15
2	106	21

a) To find the confidence interval, you must first evaluate the Wilson estimates of each proportion

$$\tilde{p}_1 = \frac{X_1 + 1}{n_1 + 2} =$$

and

$$\tilde{p}_2 = \frac{X_2 + 1}{n_2 + 2} =$$

The level C confidence interval for $p_1 - p_2$ is $(\tilde{p}_1 - \tilde{p}_2) \pm z^* SE_{\tilde{D}}$, where $\tilde{p}_1$ and $\tilde{p}_2$ were computed above and

$$SE_{\tilde{D}} = \sqrt{\frac{\tilde{p}_1(1 - \tilde{p}_1)}{n_1 + 2} + \frac{\tilde{p}_2(1 - \tilde{p}_2)}{n_2 + 2}} =$$

To complete the calculations for the confidence interval, z^* is the upper $(1 - C)/2$ standard normal critical value. What is the numerical value of C? Use it to find z^*. Compute the confidence interval in the space below and briefly summarize what the data show.

$(\tilde{p}_1 - \tilde{p}_2) \pm z^* SE_{\tilde{D}} =$

Exercise 8.45

KEY CONCEPTS - sample proportions, confidence intervals for the difference between two population proportions

a) A sample proportion is the number of successes divided by the number of observations. Suppose $\hat{p}_1$ is the proportion of farmers in favor of the program in Tippecanoe County, and $\hat{p}_2$ is the proportion in Benton County. Find the values of $\hat{p}_1$ and $\hat{p}_2$ from the information given.

b) To find the confidence interval, you must first evaluate the Wilson estimates of each proportion

$$\tilde{p}_1 = \frac{X_1 + 1}{n_1 + 2} =$$

and

$$\tilde{p}_2 = \frac{X_2 + 1}{n_2 + 2} =$$

The standard error needed to compute the confidence interval for the difference in proportions is then

$$SE_D = \sqrt{\frac{\tilde{p}_1(1 - \tilde{p}_1)}{n_1 + 2} + \frac{\tilde{p}_2(1 - \tilde{p}_2)}{n_2 + 2}} =$$

c) The level C confidence interval for p_1 - p_2 is $(\tilde{p}_1 - \tilde{p}_2) \pm z^* SE_{\tilde{D}}$ where $\tilde{p}_1$, $\tilde{p}_2$ and $SE_{\tilde{D}}$ were computed in (b), and z^* is the upper $(1 - C)/2$ standard normal critical value. What is the numerical value of C? Use it to find z^*. Now you can compute the confidence interval below.

$$(\tilde{p}_1 - \tilde{p}_2) \pm z^* SE_{\tilde{D}} =$$

Does this interval give evidence that the opinions differed in the two counties? Interpret your interval.

Exercise 8.46

KEY CONCEPTS - pooled estimate of the proportion, testing equality of two population proportions

a) The overall proportion of farmers who favor the corn checkoff program is found by combining the farmers from the two counties. It is the number in favor from both counties divided by the number sampled from both counties. If X_1 and X_2 are the number who favor the favor the program in Tippecanoe and Benton County respectively, and n_1 and n_2 are the sample sizes, use the formula below to compute the overall proportion.

$$\hat{p} = \frac{X_1 + X_2}{n_1 + n_2} =$$

b) The standard error for testing that the population proportions favoring the program is the same in two counties is expressed in terms of the overall proportion computed in (a). Evaluate the standard error using the formula below.

$$SE_{Dp} = \sqrt{\hat{p}(1-\hat{p})\left(\frac{1}{n_1} + \frac{1}{n_2}\right)} =$$

c) We are interested in seeing if there is a difference between the two counties. What does this say about the alternative? Is it one-sided or two-sided? Write down the null and alternative hypotheses in terms of the population proportions p_1 and p_2.

d) The z-statistic for testing equality of two population proportions if given by the formula

$$z = \frac{\hat{p}_1 - \hat{p}_2}{SE_{Dp}} =$$

Evaluate the z-statistic above using the value of SE_{Dp} computed in (a) and the values of $\hat{p}_1$ and $\hat{p}_2$ computed in Exercise 8.45. Compute the P-value and give your conclusion. When computing the P-value remember that the alternative is two-sided.

COMPLETE SOLUTIONS

Exercise 8.27

The Wilson estimates of each proportion are

$$\tilde{p}_1 = \frac{X_1 + 1}{n_1 + 2} = \frac{15 + 1}{84 + 2} = 0.1860$$

and

$$\tilde{p}_2 = \frac{X_2 + 1}{n_2 + 2} = \frac{21 + 1}{106 + 2} = 0.2037$$

and the standard error of the difference is

$$SE_{\tilde{D}} = \sqrt{\frac{\tilde{p}_1(1-\tilde{p}_1)}{n_1 + 2} + \frac{\tilde{p}_2(1-\tilde{p}_2)}{n_2 + 2}} = \sqrt{\frac{.1860(1-.1860)}{84 + 2} + \frac{.2037(1-.2037)}{106 + 2}}$$

$$= 0.0571$$

For the 95% confidence interval, $C = 0.95$ and we need the upper $(1 - C)/2 = (1 - 0.95)/2 = 0.025$ standard normal critical value. This can be gotten most easily from the bottom row of Table D, with $z^* = 1.96$. The confidence interval is given by

$$(\tilde{p}_1 - \tilde{p}_2) \pm z^* SE_{\tilde{D}} = (0.1860 - 0.2037) \pm 1.96(0.0571) = -0.018 \pm 0.112$$

The difference in proportions is between -0.130 and 0.094. There is no evidence of a change over time as the confidence interval includes 0, or no difference in proportions.

Exercise 8.45

a) In Tippecanoe County there were $X_1 = 263$ in favor of the program among the $n_1 = 263 + 252 = 515$ farmers sampled. (The number sampled is the number in favor plus the number opposed.) So the value of $\hat{p}_1 = 263/515 = 0.511$. Similarly, $X_1 = 260$, $n_2 = 260 + 377 = 637$, and $\hat{p}_2 = 260/637 = 0.408$.

b) The Wilson estimates of each proportion are

$$\tilde{p}_1 = \frac{X_1 + 1}{n_1 + 2} = \frac{263 + 1}{515 + 2} = 0.511$$

and

$$\tilde{p}_2 = \frac{X_2 + 1}{n_2 + 2} = \frac{260 + 1}{637 + 2} = 0.408$$

and the standard error of the difference is

$$SE_{\tilde{D}} = \sqrt{\frac{\tilde{p}_1(1 - \tilde{p}_1)}{n_1 + 2} + \frac{\tilde{p}_2(1 - \tilde{p}_2)}{n_2 + 2}} = \sqrt{\frac{.511(1 - .511)}{515 + 2} + \frac{.408(1 - .408)}{637 + 2}}$$

$$= 0.0293$$

c) For the 95% confidence interval, $C = 0.95$ and we need the upper $(1 - C)/2 = (1 - 0.95)/2 = 0.025$ standard normal critical value. This can be gotten most easily from the bottom row of Table D, with $z^* = 1.96$. The confidence interval is given by

$$(\tilde{p}_1 - \tilde{p}_2) \pm z^* SE_{\tilde{D}} = (0.511 - 0.408) \pm 1.96(0.0293) = 0.103 \pm 0.057$$
$$= (0.046, 0.160)$$

Since the interval doesn't include 0, the evidence is that the opinions differed, with more farmers in Tippecanoe in favor of the program. We estimate 10.3%

more farmers from Tippecanoe favor the program, with a 95% margin of error of 5.7%

Exercise 8.46

a) Using the data from Exercise 8.45, we have $X_1 = 263$, $X_2 = 260$, $n_1 = 263 + 252 = 515$ and $n_2 = 260 + 377 = 637$. The overall proportion is

$$\hat{p} = \frac{X_1 + X_2}{n_1 + n_2} = \frac{263 + 260}{515 + 637} = 0.454$$

b) The standard error used to test equality of the population proportions is

$$\text{SE}_{Dp} = \sqrt{\hat{p}(1 - \hat{p})\left(\frac{1}{n_1} + \frac{1}{n_2}\right)} = \sqrt{0.454(1 - 0.454)\left(\frac{1}{515} + \frac{1}{637}\right)} = 0.0295$$

c) The hypotheses are H_0: $p_1 = p_2$ and H_a: $p_1 \neq p_2$. The alternative is two-sided since we are interested in differences in either direction.

d) The numerical value of the z-statistic for testing equality of two population proportions is

$$z = \frac{\hat{p}_1 - \hat{p}_2}{\text{SE}_{Dp}} = \frac{0.511 - 0.408}{0.0295} = 3.49$$

We need to first find the probability of getting a z-statistic at least this large using Table A, and then this probability must be doubled to obtain the P-value since the alternative is two-sided.

$P(Z > 3.49) = 0.0002$ from Table A. So the P-value is 0.0004. There is very strong evidence that the proportion in favor of the program differs between the two counties. The proportion in favor of the program is higher in Tippecanoe.

CHAPTER 9

INFERENCE FOR TWO-WAY TABLES

OVERVIEW

This chapter discusses techniques for describing the relationship between two or more categorical variables. To analyze categorical variables, we use counts (frequencies) or percents (relative frequencies) of individuals that fall into various categories. **A two-way table** of such counts is used to organize data about two categorical variables. Values of the **row variable** label the rows that run across the table, and values of the **column variable** label the columns that run down the table. In each cell (intersection of a row and column) of the table, we enter the number of cases for which the row and column variables have the values (categories) corresponding to that cell.

The **row totals** and **column totals** in a **two-way table** give the marginal distributions of the two variables separately. It is usually clearest to present these distributions as percents of the table total. Marginal distributions do not give any information about the relationship between the variables. **Bar graphs** are a useful way of presenting these marginal distributions.

The conditional distributions in a two-way table help us to see relationships between two categorical variables. To find the conditional distribution of the row variable for a specific value of the column variable, look only at that one column in the table. Express each entry in the column as a percent of the column total. There is a conditional distribution of the row variable for each column in the table. Comparing these conditional distributions is one way to describe the association between the row and column variables, particularly if the column variable is the explanatory variable. When the row variable is explanatory, find the conditional distribution of the column variable for each row and compare these distributions. Side-by-side bar graphs of the conditional distributions of the row or column variable can be used to compare these distributions and describe any association that may be present.

Data on three categorical variables can be presented in a **three-way table**, printed as separate two-way tables for each value of the third variable. An association between two variables that holds for each level of this third variable can be changed, or even reversed, when the data are **aggregated** by summing over all values of the third variable. **Simpson's paradox** refers to a reversal of an association by aggregation.

There are two common models that generate data that can be summarized in a two-way table. In the first model, independent SRS's are drawn from c populations and each observation is classified according to a categorical variable that has r possible values. In the table, the c populations form the columns, and the categorical classification variable forms the rows. The null hypothesis is that the distributions of the row categorical variable are the same for each of c populations. In the second model, an SRS is drawn from a single population, and the observations are cross-classified according to two categorical variables having r and c possible values, respectively. In this model the null hypothesis is that the row and column variables are independent. When one of the variables is an explanatory variable and the other is a response, the explanatory variable is used to form the columns of the table and the response forms the rows.

A test of the null hypothesis is carried out using X^2 in both models. Although the data are generated differently, the question in both cases is quite similar: Are the distributions of the column variables the same? The cell counts are compared to the **expected cell counts** under the null hypothesis. The expected cell counts are computed using the formula

$$\text{expected count} = \frac{\text{row total} \times \text{column total}}{n}$$

where n is the total number of observations.

The **chi-square statistic** is used to test the null hypothesis by comparing the observed counts with the expected counts

$$X^2 = \sum \frac{(\text{observed - expected})^2}{\text{expected}}$$

When the null hypothesis is true, the distribution of X^2 is approximately χ^2 with $(r-1)(c-1)$ degrees of freedom. The P-value is the probability of getting differences between observed and expected counts as large as we did, and is computed as $P(\chi^2 > X^2)$. The use of the χ^2 distribution is an approximation that works well when the average expected count exceeds 5 and all of the individual expected counts are greater than 1. In the special case of the 2 x 2 table, all expected counts should exceed 5 before applying the approximation.

In addition to computing the X^2 statistic, tables or bar charts should be examined to describe the relationship between the two variables.

GUIDED SOLUTIONS

Exercise 9.1

KEY CONCEPTS - joint and marginal distributions in two-way tables

For convenience, below we reproduce the table given in the problem. The entries are in thousands of persons.

Years of school completed, by age (thousands of persons)

	Age group			
Education	25 to 34	35 to 54	55 and over	Total
Did not complete high school	5,325	9,152	16,035	30,512
Completed high school	14,061	24,070	18,320	56,451
College, 1 to 3 years	11,659	19,926	9,662	41,247
College, 4 or more years	10,342	19,878	8,005	38,225
Total	41,388	73,028	52,022	166,438

a) The joint distribution is the collection of cell proportions for the two categorical variables, where the cell proportions are the proportion each cell count is of the total number of counts in the table (in this case, 166,438 thousands of persons). For example, the cell proportion corresponding to the age group "25 to 34" and education category "Did not complete high school" is $5,325/166,438 = 0.032$.

To answer the question, you need to compute each cell proportion and enter them in the table below. This will be the joint distribution. You can double check your work by verifying that the sum of all the proportions is 1.

	Age group		
Education	25 to 34	35 to 54	55 and over
Did not complete high school	0.032		
Completed high school			
College, 1 to 3 years			
College, 4 or more years			

b) The marginal distribution of age can be found from the totals for each age group given in the bottom row of the table. Each value in this row must be divided by the total number of counts, namely 166,438, to give the proportion in each age group. Compute each of these proportions and enter the results in the table below.

	Age group		
	25 to 34	35 to 54	55 and over
Proportion			

c) The marginal distribution of education can be found from the totals for each education category given in the rightmost column of the table. Each value in this column must be divided by the total number of counts, namely 166,438, to give the proportion in each education category. Compute each of these proportions and enter the results in the table below.

Education	Proportion
Did not complete high school	
Completed high school	
College, 1 to 3 years	
College, 4 or more years	

Exercise 9.3

KEY CONCEPTS - conditional distributions and association in a two-way table

To compute the conditional distribution of age for a particular education category one must divide each cell entry in the row corresponding to the education category, by the row total (entry in rightmost column of the table). For example, for the category "Did not complete high school" the conditional distribution involves dividing the entries 5,325, 9,152, and 16,035 each by 30,512.

    Carry out these calculations and enter the results in the table below.

	Age group		
Education	25 to 34	35 to 54	55 and over
Did not complete high school			
Completed high school			
College, 1 to 3 years			
College, 4 or more years			

To summarize these results graphically, make a separate bar graph for each education category. To assist you, we have provided the axes for each plot on the following pages.

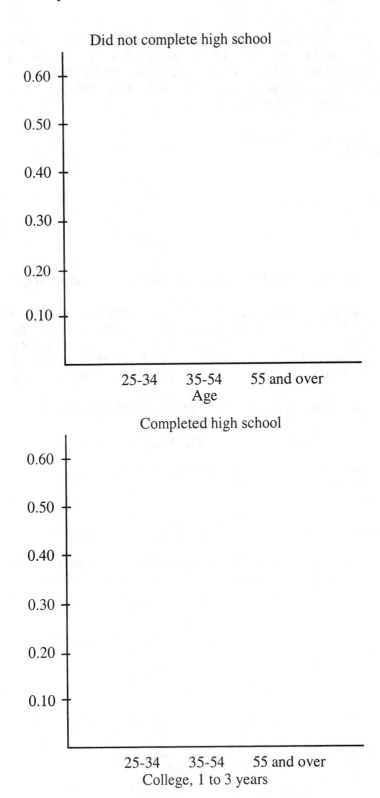

Did not complete high school

Age

Completed high school

College, 1 to 3 years

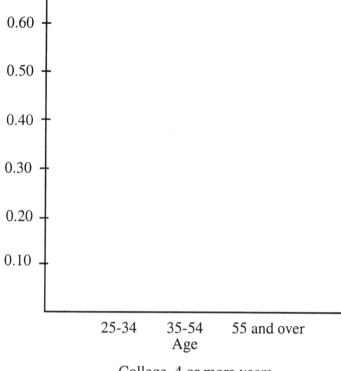

College, 4 or more years

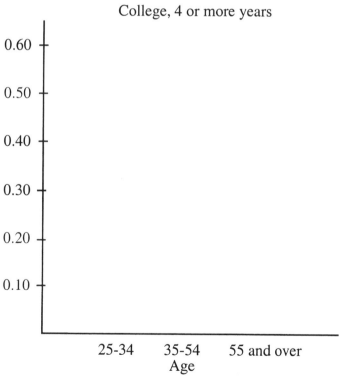

How do the distributions differ?

Exercise 9.17

KEY CONCEPTS - aggregating three-way tables, Simpson's paradox

a) Add corresponding entries in the three-way table and enter the sums in the table below.

	Admit	
Gender	Yes	No
Male		
Female		

b) Convert your table in part (a) to one involving percentages of the row totals.

	Admit	
Gender	Yes	No
Male		
Female		

c) Now repeat the type of calculations you did in part (b) for each of the original tables.

Business

	Admit	
Gender	Yes	No
Male		
Female		

Law

Gender	Admit	
	Yes	No
Male		
Female		

d) To explain the apparent contradiction observed in part (c), consider which professional school is easier to get into and which professional school males and females tend to apply to. Write your answer in plain English in the space provided. Avoid jargon and be clear!

Exercise 9.33

KEY CONCEPTS - testing independence in a 2x2 table

a) For this data, which is the explanatory variable and which is the response? In order to be an experiment, what needs to be true about the assignment of the explanatory variable to the subjects (review Sections 3.1 and 3.2)? Was this carried out here?

b) Pet ownership defines the two populations. The percentages that you compute should allow comparisons between these two populations. In this case, how does the percentage of patients alive compare for the two categories of pet ownership? Compute the appropriate percentages in the table below and state your preliminary conclusions.

Patient status	Pet ownership	
	No	Yes
Alive		
Dead		

c) In this exercise, was a single SRS drawn from a population, or were independent SRS's drawn from each population? This is important for stating the null hypothesis. Don't just state the hypotheses generally in terms of row and column variables. Be sure to state the hypotheses using the names of the variables in this particular problem.

d) The easiest way to find the value of the chi-square statistic is to use statistical software. If you don't have access to statistical software the computations must be done by hand. The first step is to determine the expected counts for each of the cells. Remember the formula is

$$\text{expected count} = \frac{\text{row total} \times \text{column total}}{n}$$

It would be easiest to add the row totals, the column totals and the value of n to the original table below before computing the expected counts. In the table below, fill in the numerical values of the expected counts below the counts in each of the cells.

Patient status	Pet ownership No	Yes	Total
Alive	28	50	
Dead	11	3	
Total			

The chi-square statistic compares the observed counts to the expected counts using the formula below. What is the value of X^2 in this problem? Remember, there are four terms in the sum, one corresponding to each of the cells.

$$X^2 = \sum \frac{(\text{observed} - \text{expected})^2}{\text{expected}} =$$

X^2 has an approximate χ^2 distribution with degrees of freedom $(r - 1)(c - 1)$, provided all the expected counts are 5 or greater. What are r and c in this case? Verify that the approximation can be applied.

Now use Table F to determine the *P*-value. It is the area to the right of the value of X^2 using the line in the table corresponding to the degrees of freedom you computed.

e) Is this observational data or an experiment? This determines the type of conclusion you can draw. Is there a cause and effect relationship or just association? Can the effects be attributed to confounding variables? Review these ideas from Sections 3.1 and 3.2 and apply them to this data set.

Exercise 9.41

KEY CONCEPTS - testing that the distributions of the response variable are the same in *c* populations

First, complete the table by filling in the row and column totals below.

Taster	Ireland	Portugal	Norway	Italy	Total
			Country		
Yes	558	345	185	402	
No	225	109	81	134	
Total					

In order to describe the data, ask yourself the following. Are the data separate simple random samples from *c* different populations with each individual classified according to a categorical variable with *r* possible values (the first model) or are they a single simple random sample from one population classified according to two categorical variables with *r* and *c* possible values, respectively (model two)? What hypothesis do you wish to test?

To test the hypothesis, we must compute the X^2 statistic, namely

$$X^2 = \sum \frac{(\text{observed} - \text{expected})^2}{\text{expected}}$$

The observed values for each cell can be read from the table. The expected counts are computed for each cell in the table using the formula

$$\text{expected count} = \frac{\text{row total} \times \text{column total}}{n}$$

where n is the total number of individuals in the table. Compute these and enter the values in the space below each of the observed counts in the following table.

		Country			
Taster	Ireland	Portugal	Norway	Italy	Total
Yes	558	345	185	402	
No	225	109	81	134	
Total					

Now compute

$$X^2 = \sum \frac{(\text{observed - expected})^2}{\text{expected}} =$$

What is the P-value? What do you conclude?

COMPLETE SOLUTIONS

Exercise 9.1

a) The joint distribution is as indicated in the completed table below.

Years of school completed, by age (thousands of persons

	Age group		
Education	25 to 34	35 to 54	55 and over
Did not complete high school	0.032	0.055	0.096
Completed high school	0.084	0.145	0.110
College, 1 to 3 years	0.070	0.120	0.058
College, 4 or more years	0.062	0.119	0.048

b) The marginal distribution of age is as in the completed table below.

	Age group		
	25 to 34	35 to 54	55 and over
Proportion	0.249	0.439	0.313

c) The marginal distribution of education is as in the completed table below.

Education	Proportion
Did not complete high school	0.183
Completed high school	0.339
College, 1 to 3 years	0.248
College, 4 or more years	0.230

Exercise 9.3

The conditional distribution of age for each of the four education categories is summarized in the rows of the table below.

	Age group		
Education	25 to 34	35 to 54	55 and over
Did not complete high school	0.175	0.300	0.526
Completed high school	0.249	0.426	0.325
College, 1 to 3 years	0.283	0.483	0.234
College, 4 or more years	0.271	0.520	0.209

Graphical displays of these conditional distributions that follow..

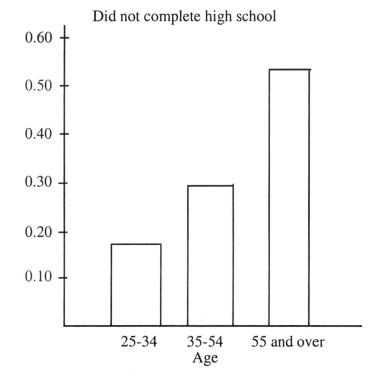

Did not complete high school

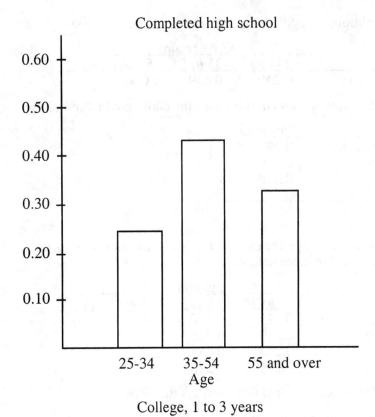

Completed high school

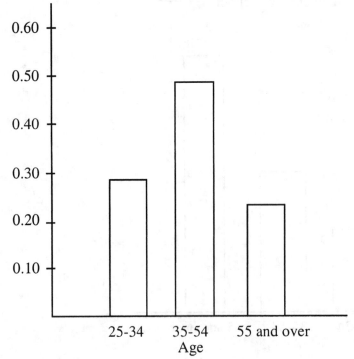

College, 1 to 3 years

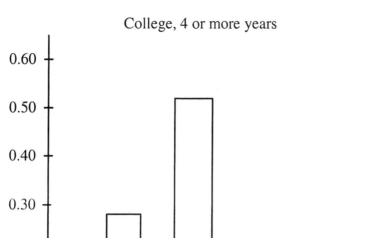

College, 4 or more years

There are several ways in which the distributions differ. First, the age distribution of those who did not complete high school differs from that of the other education categories. The proportion in each age category for those who did not complete high school increases with age, with the majority being in the 55 and over category. For the other education categories, the distributions are roughly similar to each other (and to the marginal distribution of age), with the largest proportion being in the 35 to 54 group, and the proportions in the other groups roughly similar. Second, as the amount of education increases, the proportion in the 55 and over age group decreases, and the proportion in the other age groups increases. This may suggest that extra education was less important (or less accessible) to those 55 and over than to those under 55.

Exercise 9.17

a) To make the desired table we add together the entries in the corresponding cells of the two tables given. We get

Gender	Admit	
	Yes	No
Male	490	310
Female	400	300

b) To compute the percents we must determine the row totals and then express each entry as a percentage of the row total. The row totals are 800 for the Male row and 700 for the Female row. We then obtain the following.

	Admit	
Gender	Yes	No
Male	$490/800 = 0.6125$	$310/800 = 0.3875$
Female	$400/700 = 0.5714$	$300/700 = 0.4286$

We indeed see that a higher percentage of males are admitted than females.

c) We repeat the type of calculations in (b) but applied separately to the original tables given. Note that the row totals for Business are 600 males and 300 females, and for Law are 200 males and 400 females.

	Business Admit	
Gender	Yes	No
Male	$400/600 = 0.6667$	$200/600 = 0.3333$
Female	$200/300 = 0.6667$	$100/300 = 0.3333$

	Law Admit	
Gender	Yes	No
Male	$90/200 = 0.4500$	$110/200 = 0.5500$
Female	$200/400 = 0.5000$	$200/400 = 0.5000$

Now we see that each school admits at least as high a proportion of females as males.

d) If we examine the data more closely, we see that Business tends to admit a higher percentage of students than Law. We also see that the majority of applicants to Business are male, while the majority of applicants to Law are female. The fact that more males than females apply to Business (where it is easier to get admitted) while more females than males apply to Law (where it is harder to get admitted) means that the overall admission rate for females will appear relatively low (since they are applying to Law which is hard to get into) compared to the admission rate for males (since males are applying to Business which is easy to get into).

Exercise 9.33

a) For this data, the explanatory variable is pet ownership and the response is patient status. The experimenter did not intervene and assign the subjects to own or not own a pet. They just observed this in the data. That makes this observational data, and there is a possibility that differences in patient status may be due to confounding variables, not necessarily to pet ownership.

b) Since pet ownership is the explanatory variable, we want to look at the distribution of patient status for each category of pet ownership. In this case the column percentages are of interest.

	Pet ownership	
Patient status	No	Yes
Alive	72%	94%
Dead	28%	6%
	100%	100%

There are large differences in the percentages. The death rate among those not owning pets appears to be higher.

c) The data is viewed as a sample from the population of patients with CHD who are classified according to two categorical variables. In this model, the null hypothesis is that pet ownership and patient status are independent. The alternative is that they are not independent. The alternative is general and does not imply any particular type of dependence or relationship between the variables.

d) The original table is reproduced below with the expected counts for each cell listed below each of the counts. The column totals, row totals and n have been added to the table as well. The expected counts were computed as follows:

(Alive, No) $\text{expected count} = \dfrac{\text{row total} \times \text{column total}}{n} = \dfrac{78 \times 39}{92} = 33.07$

(Alive, Yes) $\text{expected count} = \dfrac{78 \times 53}{92} = 44.93$

(Dead, No) $\text{expected count} = \dfrac{14 \times 39}{92} = 5.93$

(Dead, Yes) $\text{expected count} = \dfrac{14 \times 53}{92} = 8.07$

	Pet ownership		
Patient status	No	Yes	
Alive	28	50	78
	33.07	44.93	
Dead	11	3	14
	5.93	8.07	
	39	53	92

The numerical value of X^2 is computed using the formula

$$X^2 = \sum \frac{(\text{observed} - \text{expected})^2}{\text{expected}}$$

$$= \frac{(28 - 33.07)^2}{33.07} + \frac{(50 - 44.93)^2}{44.93} + \frac{(11 - 5.93)^2}{5.93} + \frac{(3 - 8.07)^2}{8.07}$$

$$= 0.777 + 0.572 + 4.33 + 3.19 = 8.87$$

Since r and c are both 2, the degrees of freedom are $(2 - 1)(2 - 1) = 1$. All the expected counts exceed 5 so the χ^2 approximation can be used. From the first row of Table F (corresponding to 1 df) we see that the P-value is between 0.0025 and 0.005. (Statistical software gives $P(\chi^2 > 8.87) = 0.003$ as the P-value).

e) Since this is observational data, you need to make sure that when you state your conclusions you do not imply that there is a cause and effect relationship. In this case pet ownership and patient status are associated. One-year survival rates are higher for those who own a pet. But remember there could be confounding variables. For example, perhaps those who own pets may tend to have different lifestyles. It's possible they have more activities and interests than those who do not own pets and this is what is helping to improve their survival. Certainly, walking your dog may be helpful in improving your survival rate if you have CHD, and dog owners may be more likely to take walks (of course, taking a walk without a dog may be just as beneficial).

This study didn't show that buying a pet upon your discharge from the coronary care unit would improve your survival. In order to show this, those leaving the coronary unit would need to be divided into two groups at random. Those in the first group would be given pets and those in the second would not be given pets. The survival rates of the two groups after one year would then be compared. Of course, while this experiment would show cause and effect, there are obvious practical difficulties in carrying it out.

Exercise 9.41

The completed table is given below.

Taster	Ireland	Portugal	Norway	Italy	Total
Yes	558	345	185	402	1490
No	225	109	81	134	549
Total	783	454	266	536	2039

Country

The data consist of samples of people from different populations (the four countries) in the table. The individuals in each sample are categorized as being able to taste PTC (yes) or not being able to taste PTC (no). We assume the samples are random samples (we are not told this is the case, and if they cannot be viewed as random samples the analysis below may be inappropriate). In this case, we have the first model and we are testing whether the distributions of a categorical response variable with $r = 2$ possible values (yes or no) are the same in each of $c = 4$ populations.

The observed and expected counts are as follows.

$n = 2039$

(Yes, Ireland): $\quad$ expected count $= \dfrac{1490 \times 783}{2039} = 572.18$

(No, Ireland): $\quad$ expected count $= \dfrac{549 \times 783}{2039} = 210.82$

(Yes, Portugal): $\quad$ expected count $= \dfrac{1490 \times 454}{2039} = 331.76$

(No, Portugal): $\quad$ expected count $= \dfrac{549 \times 454}{2039} = 122.24$

(Yes, Norway): $\quad$ expected count $= \dfrac{1490 \times 266}{2039} = 194.38$

(No, Norway): $\quad$ expected count $= \dfrac{549 \times 266}{2039} = 71.62$

(Yes, Italy): $\quad$ expected count $= \dfrac{1490 \times 536}{2039} = 391.68$

(No, Italy): $\quad$ expected count $= \dfrac{549 \times 536}{2039} = 144.32$

Taster	Ireland	Country Portugal	Norway	Italy	Total
Yes	558	345	185	402	1490
	572.18	331.76	194.38	391.68	
No	225	109	81	134	549
	210.82	122.24	71.62	144.32	
Total	783	454	266	536	2039

We now compute

$$X^2 = \sum \frac{(\text{observed} - \text{expected})^2}{\text{expected}} = \frac{(558 - 572.18)^2}{572.18} + \frac{(225 - 210.82)^2}{210.82} +$$

$$\frac{(345 - 331.76)^2}{331.76} + \frac{(109 - 122.24)^2}{122.24} + \frac{(185 - 194.38)^2}{194.38} +$$

$$\frac{(81 - 71.62)^2}{71.62} + \frac{(402 - 391.68)^2}{391.68} + \frac{(134 - 144.32)^2}{144.32}$$

$$= 0.35 + 0.95 + 0.53 + 1.43 + 0.45 + 1.23 + 0.27 + 0.74$$

$$= 5.95.$$

$X^2 = 5.95$ and has $(r-1)(c-1) = (2-1)(4-1) = 3$ degrees of freedom. Going to Table F and using the line corresponding to 3 degrees of freedom, we find this lies between the entries are 5.32, corresponding to an upper tail area of 0.15, and 6.25, corresponding to an upper tail probability of 0.10. We conclude the P-value = $P(\chi^2 > 5.95)$ is between 0.15 and 0.10. Thus, there is not strong evidence against the null hypothesis that the distributions of the categorical response variables (yes can taste PTC or no cannot taste PTC) are the same in each of four countries. In other words, the proportion of PTC tasters does not vary among the four countries.

Note: If you did this using software you should get comparable results to the hand calculations presented above.

CHAPTER 10

INFERENCE FOR REGRESSION

SECTIONS 10.1 and 10.2

OVERVIEW

The statistical model for **simple linear regression** is

$$y_i = \beta_0 + \beta_1 x_i + \varepsilon_i$$

where $i = 1, 2,..., n$. The deviations ε_i are assumed to be independent and normally distributed with mean 0 and standard deviation σ. The **parameters** of the model are the intercept β_0, the slope β_1, and σ. β_0 and β_1 are estimated by the slope b_0 and intercept b_1 of the **least-squares regression line**. Given n observations on an explanatory variable x and a response variable y,

$$(x_1, y_1), (x_2, y_2), ..., (x_n, y_n)$$

recall that the formula for the slope and intercept of the least-squares regression line are

$$b_1 = r \frac{s_y}{s_x}$$

and

$$b_0 = \bar{y} - b_1 \bar{x}$$

where r is the correlation between y and x, $\bar{y}$ is the mean of the y observations, s_y is the standard deviation of the y observations, $\bar{x}$ is the mean of the x

observations, and s_x is the standard deviation of the x observations. The standard deviation σ is estimated by

$$s = \sqrt{\frac{\sum e_i^2}{n-2}}$$

where the e_i are the **residuals**

$$e_i = y_i - \hat{y}_i$$

and

$$\hat{y}_i = b_0 + b_1 x_i$$

b_0, b_1, and s are usually calculated using a calculator or statistical software.

A **level C confidence interval** for β_1 is

$$b_1 \pm t^* SE_{b_1}$$

where t^* is the upper $(1-C)/2$ critical value for the $t(n-2)$ distribution and

$$SE_{b_1} = \frac{s}{\sqrt{\sum (x_i - \bar{x})^2}}$$

is the standard error of the slope b_1. SE_{b_1} is usually computed using a calculator or statistical software. (The formula above is actually given in Section 10.2 of the text, but reproduced here for continuity of exposition. If your class is not covering Section 10.2, you need not worry about the formula.) The **test of the hypothesis $H_0: \beta_1 = 0$** is based on the t statistic

$$t = \frac{b_1}{SE_{b_1}}$$

with P-values computed from the $t(n-2)$ distribution. There are similar formulas for confidence intervals and tests for β_0, but using the standard error of the intercept b_0

$$SE_{b_0} = s \sqrt{\frac{1}{n} + \frac{\bar{x}^2}{\sum (x_i - \bar{x})^2}}$$

in place of SE_{b_1}. As with SE_{b_1}, SE_{b_0} is usually computed using a calculator or statistical software. (The formula above is actually given in Section 10.2 of the text, but reproduced here for continuity of exposition. If your class is not covering Section 10.2, you need not worry about the formula.) Note that inferences for the intercept are meaningful only in special cases.

The **estimated mean response** for the subpopulation corresponding to the value x^* of the explanatory variable is

$$\hat{\mu}_y = b_0 + b_1 x^*$$

A **level C confidence interval for the mean response** is

$$\hat{\mu}_y \pm t^* \mathrm{SE}_{\hat{\mu}}$$

where t^* is the upper $(1-C)/2$ critical value for the $t(n-2)$ distribution and

$$\mathrm{SE}_{\hat{\mu}} = s \sqrt{\frac{1}{n} + \frac{(x^* - \bar{x})^2}{\sum (x_i - \bar{x})^2}}$$

$\mathrm{SE}_{\hat{\mu}}$ is usually computed using a calculator or statistical software. (The formula above is actually given in Section 10.2 of the text, but reproduced here for continuity of exposition. If your class is not covering Section 10.2, you need not worry about the formula.)

The **estimated value of the response variable** y for a future observation from the subpopulation corresponding to the value x^* of the explanatory variable is

$$\hat{y} = b_0 + b_1 x^*$$

A **level C prediction interval** for the estimated response is

$$\hat{y} \pm t^* \mathrm{SE}_{\hat{y}}$$

where t^* is the upper $(1-C)/2$ critical value for the $t(n-2)$ distribution and

$$\mathrm{SE}_{\hat{y}} = s \sqrt{1 + \frac{1}{n} + \frac{(x^* - \bar{x})^2}{\sum (x_i - \bar{x})^2}}$$

$\mathrm{SE}_{\hat{y}}$ is usually computed using a calculator or statistical software. (The formula above is actually given in Section 10.2 of the text, but reproduced here for continuity of exposition. If your class is not covering Section 10.2, you need not worry about the formula.)

The **ANOVA table** for a linear regression gives the total sum of squares SSM for the model, the total sum of squares SSE for error, the total sum of squares SST for all sources of variation, the degrees of freedom DFM, DFE, and DFT for these sum of squares, the mean square MSM for the model, and the mean square MSE for error. The total sum of squares are

$$\mathrm{SSM} = \sum (\hat{y}_i - \bar{y})^2$$

$$\mathrm{SSE} = \sum (y_i - \hat{y}_i)^2$$

$$\text{SST} = \sum (y_i - \bar{y})^2$$

and the degrees of freedom are

$$\text{DFM} = 1$$

$$\text{DFE} = n - 2$$

$$\text{DFT} = n - 1$$

The mean sum of squares MS are defined by the relation

$$\text{MS} = \frac{\text{sum of squares}}{\text{degrees of freedom}}$$

The ANOVA table usually has a form like the following.

Source	Degrees of freedom	Sum of Squares	Mean Square	F
Model	DFM	SSM	MSM = SSM/DFM	MSM/MSE
Error	DFE	SSE	MSE = SSE/DFE	
Total	DFT	SST		

The **ANOVA F statistic** is the ratio MSM/MSE and is used to test $H_0: \beta_1 = 0$, versus the two-sided alternative. Under H_0, this statistic has an $F(1, n - 2)$ distribution which can be used to compute P-values using Table E.

When the variables y and x are jointly normal, the sample correlation is an estimate of the population correlation ρ. The test of $H_0: \rho = 0$ is based on the **t statistic**

$$t = \frac{r\sqrt{n-2}}{\sqrt{1-r^2}}$$

which has a $t(n - 2)$ distribution under H_0. This test statistic is numerically identical to the t statistic used to test $H_0: \beta_1 = 0$.

The **square of the sample correlation** can be written as

$$r^2 = \text{SSM/SST}$$

and is interpreted as the proportion of the variability in the response variable y that is explained by the explanatory variable x in the simple linear regression.

SAMPLE PROBLEMS

GUIDED SOLUTIONS

Exercise 10.20

KEY CONCEPTS - scatterplots, linear regression, significance tests for the slope, prediction intervals

a) What is the explanatory variable and what is the response variable here? Remember that in the scatterplot, the horizontal axis represents the explanatory variable and the vertical axis the response variable. Sketch your scatterplot in the space provided below. Are there any outliers or unusual observations in the plot?

b) If possible, use statistical software to compute the equation of the least-squares line. If you do not have access to statistical software, let x be heart rate HR and y oxygen uptake V02, and compute the following sample quantities from the data.

$$s_x =$$
$$\bar{x} =$$
$$s_y =$$
$$\bar{y} =$$
$$r = \text{the correlation between } x \text{ and } y =$$

From these sample quantities, compute

$$b_1 = r\frac{s_y}{s_x} =$$

and

$$b_0 = \bar{y} - b_1\bar{x} =$$

c) We wish to test the hypothesis H_0: $\beta_1 = 0$. Should H_a be one- or two-sided? Is there anything in the problem suggesting that we are interested in showing the slope is positive? Negative? Or are we just interested in whether or not the slope is 0?

The actual test is best done using statistical software. If you do not have access to statistical software, you will need to compute the residuals $e_i = y_i - \hat{y}_i$ for the data, then compute

$$s = \sqrt{\frac{\sum e_i^2}{n-2}} =$$

and

$$SE_{b_1} = \frac{s}{\sqrt{(x_i - \bar{x})^2}} = \frac{s}{\sqrt{(n-1)s_x^2}} =$$

and finally

$$t = \frac{b_1}{SE_{b_1}} =$$

Use Table D (critical values of the t distribution) to determine the P-value. What are the appropriate degrees of freedom? Explain in plain English what you would conclude from this test.

d) These 95% prediction intervals are most easily computed using statistical software. Minitab, for example, will compute prediction intervals for you. Write these intervals below.

To compute these prediction intervals by hand, we recall the relevant formulas. To compute a 95% prediction interval for the estimated response at the value x^* of the explanatory variable you will need to compute the following

$$\hat{y} \pm t^* SE_{\hat{y}}$$

where

$$\hat{y} = b_0 + b_1 x^*$$

and t^* is the upper 0.025 critical value for the $t(n - 2) = t(17)$ distribution (which from Table D is 2.11) and

$$SE_{\hat{y}} = s\sqrt{1 + \frac{1}{n} + \frac{(x^* - \bar{x})^2}{\sum(x_i - \bar{x})^2}}$$

b_0, b_1, and s have already been computed in (b) and (c). Also notice that in the formula for $SE_{\hat{y}}$, $\sum(x_i - \bar{x})^2 = (n - 1)s_x^2$. To further clarify how the computations go, we give the calculations for the case $x^* = 96$.

Recall from (b) that $b_1 = 0.03866$ and $b_0 = -2.805$, and from (c) $s = 0.1205$. We find for $x^* = 96$

$$\hat{y} = b_0 + b_1 x^* = -2.805 + 0.03866(96) = 0.9064$$

$$SE_{\hat{y}} = s\sqrt{1 + \frac{1}{n} + \frac{(x^* - \bar{x})^2}{\sum(x_i - \bar{x})^2}} = s\sqrt{1 + \frac{1}{n} + \frac{(x^* - \bar{x})^2}{(n-1)s_x^2}}$$

$$= (0.1205)\sqrt{1 + \frac{1}{19} + \frac{(96 - 107)^2}{(18)(11.8275)^2}}$$

$$= (0.1205)1.0491$$

$$= 0.1264$$

Now the upper 0.025 critical value for the $t(n - 2) = t(17)$ distribution from Table D is $t^* = 2.11$ so our 95% prediction interval is

$$\hat{y} \pm t^* SE_{\hat{y}} = 0.9064 \pm 2.11(0.1264) = 0.9064 \pm 0.2667$$

Now try the calculations for $x^* = 115$ on your own.

e) In determining whether the researchers should use the predicted V02 in place of the measured V02 for this individual consider the following. From the scatterplot in (a), how close are the points to lying perfectly on a straight line? What does the P-value of the test in (c) indicate about using the least-squares regression line for prediction? What is the margin of error for the 95% prediction intervals found in (d)? Do the observed values of V02 for values of HR near 96 and 115 fall inside or outside the prediction intervals? Should the answer depend on the margin of error for prediction that the researchers can tolerate?

Exercise 10.32

KEY CONCEPTS - the ANOVA table, the ANOVA F statistic, the square of the correlation

a) The ANOVA table is most easily computed using statistical software. Copy your ANOVA table in the space below.

Source	Degrees of freedom	Sum of Squares	Mean Square	F
Model				
Error				
Total				

To compute the sums of squares by hand, recall the relevant formulas.

$$\text{Sum of squares for the model} = \text{SSM} = \sum (\hat{y}_i - \bar{y})^2$$

$$\text{Sum of squares for error} = \text{SSE} = \sum (y_i - \hat{y}_i)^2$$

$$\text{Sum of squares total} = \text{SST} = \sum (y_i - \bar{y})^2 = (n-1) s_y^2$$

To simplify things, we note that

$$\text{SSE} = \sum (y_i - \hat{y}_i)^2 = (n-2)s^2 =$$

$$SST = \sum (y_i - \bar{y})^2 = (n-1)s_y^2 =$$

where s_y and s were computed in (b) of Exercise 10.20. One then can compute

SSM = SST - SSE =

You will need to compute the degrees of freedom

Degrees of freedom for SSM = DFM = 1

Degrees of freedom for SSE = DFE = $n - 2 =$

Degrees of freedom for SST = DFT = $n - 1 =$

Finally, the mean sum of squares MS are calculated from the relation

$$MS = \frac{\text{sum of squares}}{\text{degrees of freedom}}$$

and F = MSM/MSE =

After computing these quantities, fill in the ANOVA table on the previous page.

b) What parameter in the linear regression model is tested by the ANOVA F statistic? If you are not sure, you may wish to refer to the Chapter Overview to refresh your memory as to what null hypothesis is tested by the ANOVA F statistic.

Interpret this hypothesis in terms of the relation between HR and V02.

c) Recall, the ANOVA F statistic is the ratio MSM/MSE and you computed its value in (a). It is used to test $H_0: \beta_1 = 0$, versus the two-sided alternative. Under H_0, this statistic has an $F(1, n-2)$ distribution. What is n here? P-values are computed using Table E (refer to Chapter 7 for more on the F-distribution).

d) Recall that the t statistic calculated in Exercise 10.20 was $t = 16.1$. Refer to (a) for the value of F.

e) Recall that the square of the sample correlation can be written as

$$r^2 = SSM/SST$$

and is interpreted as the proportion of the variation in the response variable y (oxygen uptake here) that is explained by the explanatory variable x (heart rate here) in the simple linear regression. Now compute r^2.

Exercise 10.33

KEY CONCEPTS - inferences for the correlation r

a) Recall that when the variables y and x are jointly normal, the sample correlation r is an estimate of the population correlation ρ. The test of H_0: $\rho = 0$ is based on the t statistic

$$t = \frac{r\sqrt{n-2}}{\sqrt{1-r^2}} =$$

Complete the calculation of t.

b) If higher birth weights are associated with higher incomes, is the correlation positive or negative? Is the alternative, therefore, one-sided or two-sided? State H_a in terms of ρ.

H_a:

c) Recall that the t statistic in (a) has a $t(n - 2)$ distribution under H_0. You can use Table D to estimate the P-value. Remember the form of the alternative hypothesis as you compute this P-value.

What do you conclude?

COMPLETE SOLUTIONS

Exercise 10.20

a) Since we are interested in whether HR can be used to predict V02, we treat HR as the explanatory variable and V02 as the response. A scatterplot of HR and V02 is given below.

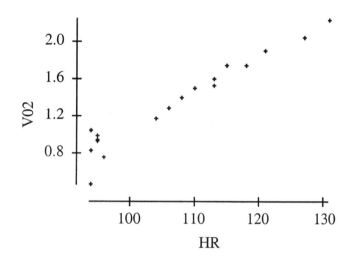

There are no outliers or other unusual points in the plot.

b) Using statistical software, we find the equation of the least-squares regression line is

$$HR = -2.8044 + 0.038652 V02$$

For those doing the calculations by hand, we find the following values of the sample statistics

$$s_x = 11.8275$$
$$\bar{x} = 107$$
$$s_y = 0.4719$$
$$\bar{y} = 1.33142$$
$$r = \text{the correlation between } x \text{ and } y = 0.969$$

From these sample quantities, compute

$$b_1 = r\frac{s_y}{s_x} = (0.969)\frac{0.4719}{11.8275} = 0.03866$$

and

$$b_0 = \bar{y} - b_1\bar{x} = 1.33142 - (0.03866)(107) = -2.805$$

which, except for round-off error, agrees with the values found with statistical software.

c) In this problem we are simply interested in determining whether or not the slope is 0. This suggests that we should test the hypotheses

$$H_0: \beta_1 = 0$$

$$H_a: \beta_1 \neq 0$$

This test would be based on the test statistic $t = \dfrac{b_1}{\mathrm{SE}_{b_1}}$. Using statistical software, we find $t = 16.1$ and that the P-value of our (two-sided) test is less than 0.001, which is obtained from the critical values of the $t(n - 2) = t(17)$ distribution in Table D. Because the test is two-sided, we must double the tail probability given in the Table to get the P-value.

If you do the calculations by hand, compute the residuals $e_i = y_i - \hat{y}_i$ for the data. Then one finds

$$s = \sqrt{\frac{\sum e_i^2}{n-2}} = \sqrt{\frac{0.2468}{17}} = 0.1205$$

and

$$\mathrm{SE}_{b_1} = \frac{s}{\sqrt{(x_i - \bar{x})^2}} = \frac{s}{\sqrt{(n-1)s_x^2}} = \frac{0.1205}{\sqrt{(18)(11.8275)^2}} = 0.0024$$

and finally

$$t = \frac{b_1}{\mathrm{SE}_{b_1}} = \frac{0.03866}{0.0024} = 16.1$$

Use Table D (critical values of the t distribution) to determine the P-value. Here we are interested in the $t(n - 2) = t(17)$ distribution and we find the P-value is less than 0.001 (we must double the tail probability in Table D since we have a two-sided hypothesis).

We conclude that there is strong evidence against the null hypothesis that there is not a straight line relation between HR and V02. However, even though the P-value is very small, this test does not tell us if the equation of the least-squares regression line will yield accurate predictions of V02 from the values of HR. All we can really conclude is that predictions using the equation of the least-squares regression line will be much better than simply predicting V02 to be the mean of the observed values, regardless of the value of HR. The issue of the accuracy of the prediction using the least-squares regression line is further explored in (d).

d) Using statistical software we find the following.

> For HR = 96 we predict V02 to be 0.9062 and a 95% prediction interval for V02 is (0.6396, 1.1729) or 0.9062 ± 0.2667.

> For HR = 115 we predict V02 to be 1.6406 and a 95% prediction interval for V02 is (1.3767, 1.9046) or 1.6406 ± 0.2640.

For those doing the calculations by hand, the 95% prediction interval for $x^* = 95$ was given in the Guided Solutions. For $x^* = 115$ we find

$$\hat{y} = b_0 + b_1 x^* = -2.805 + 0.03866(115) = 1.6409$$

$$SE_{\hat{y}} = s\sqrt{1 + \frac{1}{n} + \frac{(x^* - \bar{x})^2}{\sum(x_i - \bar{x})^2}} = s\sqrt{1 + \frac{1}{n} + \frac{(x^* - \bar{x})^2}{(n-1)s_x^2}}$$

$$= (0.1205)\sqrt{1 + \frac{1}{19} + \frac{(115 - 107)^2}{(18)(11.8275)^2}}$$

$$= (0.1205)1.0383$$

$$= 0.1251$$

Now the upper 0.025 critical value for the $t(n - 2) = t(17)$ distribution from Table D is $t^* = 2.11$ so our 95% prediction interval is

$$\hat{y} \pm t^* SE_{\hat{y}} = 1.6409 \pm 2.11(0.1251) = 1.6409 \pm 0.2640$$

e) Although from (c) we saw that the P-value of the test of the hypotheses

$$H_0: \beta_1 = 0$$

$$H_a: \beta_1 \neq 0$$

is very small, this test does not tell us if the equation of the least-squares regression line will yield accurate predictions of V02 from the values of HR. The scatterplot in (a) shows that the straight line relation is good but far from perfect, i.e., there is some variation of the points around a straight line. This is further supported by the results of (d). For HR = 96 we predict V02 to be 0.9064 (which is above the observed value at HR = 96) and a 95% prediction interval for V02 is 0.9064 ± 0.2667 or (0.6397, 1.1731). Nearly all the observed values for HR = 94, 95, and 96 fall in this interval. For HR = 115 we predict V02 to be 1.6409 and a 95% prediction interval for V02 is 1.6409 ± 0.2640 or (1.3769, 1.9049). All the observed values for HR =108 to HR = 121 fall in this interval. It would appear that predictions using the least-squares regression line are not useful for distinguishing between changes in HR of 1 or 2, but are useful for distinguishing between values of HR that differ by about 10. Ultimately, whether the researchers should use the predicted V02 in place of the measured V02 depends on how accurately the value of V02 needs to be known. The 95% prediction intervals have margins of error of about 0.26. Is such a margin of error too large or is it adequate for the purposes of the researchers? This is a question the researchers, not the statistician, need to answer.

Exercise 10.32

a) From statistical software we get the following ANOVA table.

Source	Degrees of freedom	Sum of Squares	Mean Square	F
Model	1	3.7619	3.7619	259.27
Error	17	0.2467	0.0145	
Total	18	4.0085		

If the calculations are done by hand we find

$$\text{SSE} = \sum (y_i - \hat{y}_i)^2 = (n-2)s^2 = (17)(0.1205)^2 = 0.2468$$

$$\text{SST} = \sum (y_i - \bar{y})^2 = (n-1)s_y^2 = (18)(0.4719)^2 = 4.0084$$

$$\text{SSM} = \text{SST} - \text{SSE} = 4.0084 - 0.2468 = 3.7616$$

$$\text{Degrees of freedom for SSM} = \text{DFM} = 1$$

$$\text{Degrees of freedom for SSE} = \text{DFE} = n - 2 = 17$$

$$\text{Degrees of freedom for SST} = \text{DFT} = n - 1 = 18$$

$$\text{MSM} = \text{SSM/DFM} = 3.7616/1 = 3.7616$$

$$\text{MSE} = \text{SSE/DFE} = 0.2468/17 = 0.0145$$

and

$$F = \text{MSM/MSE} = 3.7616/0.0145 = 259.42$$

Except for round-off error, this agrees with the values found using computer software.

b) The ANOVA F statistic is the ratio MSM/MSE and is used to test the null hypothesis H_0: $\beta_1 = 0$, where β_1 is the slope of the regression line in our linear regression model. In practical terms, this tests whether there is any evidence of a straight line relation between HR and V02.

c) Under H_0, the F statistic has an $F(1, n - 2) = F(1, 17)$ distribution. In (a) we saw that here the value of the F statistic is 259.27. From Table E, referring to the column labeled by 1 and the row labeled by 17, we see that the P-value is below 0.001.

d) $t^2 = (16.1)^2 = 259.21$. The value of F is 259.27. These agree, up to round-off error.

e) From our ANOVA table we find

$$r^2 = \text{SSM/SST} = 3.7619/4.0085 = 0.938$$

Exercise 10.33

a) Here $r = 0.39$ and $n = 40$. Thus

$$t = \frac{r\sqrt{n-2}}{\sqrt{1-r^2}} = \frac{0.39\sqrt{40-2}}{\sqrt{1-.39^2}} = \frac{0.39(6.16)}{\sqrt{0.8479}} = 2.61$$

b) If higher birth weights are expected to be associated with higher incomes, the correlation is expected to be positive. Because this is what the researchers wish to demonstrate, the alternative hypothesis is

$$H_a\text{: } \rho > 0$$

c) The t statistic in (a) has a $t(n - 2) = t(38)$ distribution under H_0. Using Table D to estimate the P-value we find the P-value is less than 0.0045 (we use the upper tail probability in Table D because we have a one-sided hypothesis of the form H_a: $\rho > 0$). We would conclude that there is strong evidence that population correlation ρ is greater than 0. Simply stating that the null hypothesis is rejected does not tell us anything about how different ρ is from 0.

CHAPTER 11

MULTIPLE REGRESSION

OVERVIEW

Multiple linear regression extends the techniques of simple linear regression to situations involving $p > 1$ explanatory variables $x_1, x_2, \ldots, x_p$. The statistical model for multiple linear regression is

$$y_i = \beta_0 + \beta_1 x_{i1} + \beta_2 x_{i2} + \ldots + \beta_p x_{ip} + \varepsilon_i$$

where $i = 1, 2, \ldots, n$.. The deviations ε_i are assumed to be independent and normally distributed with mean 0 and standard deviation σ. The **parameters** of the model are $\beta_0, \beta_1, \beta_2, \ldots, \beta_p$, and σ. The β's are estimated by $b_0, b_1, b_2, \ldots, b_p$ by the **principle of least-squares**. σ is estimated by

$$s = \sqrt{MSE} = \frac{\sum e_i^2}{n - p - 1}$$

where the e_i are the **residuals**

$$e_i = y_i - \hat{y}_i$$

and

$$\hat{y}_i = b_0 + b_1 x_{i1} + b_2 x_{i2} + \ldots + b_p x_{ip}$$

In practice, the b's and s are calculated using statistical software.

A **level C confidence interval** for β_j is

$$b_j \pm t^* \mathrm{SE}_{b_j}$$

where t^* is the upper $(1-C)/2$ critical value for the $t(n - p - 1)$ distribution. SE_{b_j} is the standard error of b_j and in practice is computed using statistical software.

The **test of the hypothesis** H_0: $\beta_j = 0$ is based on the t **statistic**

$$t = \frac{b_j}{\mathrm{SE}_{b_j}}$$

with P-values computed from the $t(n - p - 1)$ distribution. In practice statistical software is used to carry out these tests.

In multiple regression, interpretation of these confidence intervals and tests depends on the particular explanatory variables in the multiple regression model. The estimate of β_j represents the effect of the explanatory variable x_j when it is added to a model already containing the other explanatory variables.

The test of H_0: $\beta_j = 0$ tells us if the improvement in the ability of our model to predict the response y by adding x_j to a model already containing the other explanatory variables is statistically significant. It does not tell us if x_j would be useful for predicting the response in multiple regression models with a different collection of explanatory variables.

The **ANOVA table** for a multiple regression is analogous to that in simple linear regression. It gives the sum of squares SSM for the model, the sum of squares SSE for error, the total sum of squares SST for all sources of variation, the degrees of freedom DFM, DFE, and DFT for these sums of squares, the mean square MSM for the model, and the mean squares MSE for error. The sums of square are

$$\mathrm{SSM} = \sum (\hat{y}_i - \bar{y})^2$$

$$\mathrm{SSE} = \sum (y_i - \hat{y}_i)^2$$

$$\mathrm{SST} = \sum (y_i - \bar{y})^2$$

and the degrees of freedom are

$$\mathrm{DFM} = p$$
$$\mathrm{DFE} = n - p - 1$$
$$\mathrm{DFT} = n - 1$$

The mean squares MS are defined by the relation

$$MS = \frac{\text{sum of squares}}{\text{degrees of freedom}}$$

These quantities are computed using statistical software in practice. The results are often summarized in an ANOVA table which usually has a form like the following.

Source	Degrees of freedom	Sum of Squares	Mean Square	F
Model	DFM	SSM	SSM/DFM	MSM/MSE
Error	DFE	SSE	SSE/DFE	
Total	DFT	SST		

The **ANOVA F statistic** is the ratio MSM/MSE and is used to test

$$H_0: \beta_1 = \beta_2 = \ldots = \beta_p = 0$$

Under H_0, this statistic has an $F(p, n - p - 1)$ distribution which can be used to compute P-values using Table E. Notice that evidence against H_0 only tells us that at least one of the β_j differs from 0 but not which one. Deciding which $\beta_j \neq 0$ requires further analysis using the procedures for confidence intervals or hypothesis tests for the individual β_j mentioned above, keeping in mind the difficulties of interpreting these individual inferences.

The **squared multiple correlation** can be written as

$$R^2 = \text{SSM/SST}$$

and is interpreted as the proportion of the variability in the response variable y that is explained by the explanatory variables $x_1, x_2, \ldots, x_p$ in the multiple regression.

SAMPLE PROBLEMS

GUIDED SOLUTIONS

Exercise 11.44

KEY CONCEPTS - simple linear regression, normal quantile plots, residuals plots

a) Summarize the results of your analysis by filling in the tables below. We recommend that the analysis be carried out using statistical software.

Source	Degrees of freedom	Sum of Squares	Mean Square	F
Model				
Error				
Total				

The regression equation is

$$\text{Corn} = \underline{\hspace{1cm}} + \underline{\hspace{1cm}} \text{Year}$$

Variable	Parameter Estimate	Std. Error	*t*-ratio	*P*-value
Constant				
Year				

$s = \sqrt{\text{MSE}} = \underline{\hspace{1.5cm}}$ $R^2 = \underline{\hspace{1.5cm}}$

What do you conclude about the significance test for the slope?

b) Use statistical software to make a normal quantile plot. Is there any indication in the plot that suggests that the residuals are not normal?

c) Use statistical software to make a plot of the residuals versus soybean yield. Does the plot indicate that soybean yield might be useful in a multiple linear regression model with year to predict corn yield?

Exercise 11.45

KEY CONCEPTS - simple linear regression, normal quantile plots, residuals plots

a) Summarize the results of your analysis by filling in the tables below. We recommend that the analysis be carried out using statistical software.

Source	Degrees of freedom	Sum of Squares	Mean Square	F
Model				
Error				
Total				

The regression equation is

$$\text{Corn} = \underline{\hspace{1cm}} + \underline{\hspace{1cm}}\text{Soybeans}$$

Variable	Parameter Estimate	Std. Error	t-ratio	P-value
Constant				
Soybeans				

$$s = \sqrt{\text{MSE}} = \underline{\hspace{2cm}} \qquad R^2 = \underline{\hspace{2cm}}$$

What do you conclude about the significance test for the slope?

b) Use statistical software to make a normal quantile plot. Is there any indication in the plot that suggests that the residuals are not normal?

c) Use statistical software to make a plot of the residuals versus year. Does the plot indicate that year might be useful in a multiple linear regression model with soybean yield to predict corn yield?

Exercise 11.46

KEY CONCEPTS - multiple linear regression, R^2, interpretation of the regression coefficients in a multiple linear regression, plots of residuals

a) Summarize the results of your analysis by filling in the ANOVA table below. We recommend that the analysis be carried out using statistical software.

Source	Degrees of freedom	Sum of Squares	Mean Square	F
Model				
Error				
Total				

$s = \sqrt{MSE} = \underline{\hspace{1.5cm}}$ $R^2 = \underline{\hspace{1.5cm}}$

The null and alternative hypotheses tested by the ANOVA F test are

H_0:

H_a:

The P-value of your test of H_0 is $\underline{\hspace{3cm}}$ (you may be able to read this directly from the statistical software package you use).

What do you conclude?

b) Remember, R^2 = SSM/SST is interpreted as the proportion of the variability in the response variable y that is explained by the explanatory variables x_1, $x_2,\ldots, x_p$ in the multiple regression. What is the value of R^2 here? (You may be able to read this directly from the statistical software package you use.) Compare this with the values found in the previous two exercises (for the values of R^2 in these exercises, see their complete solutions).

c) From the results you obtained with your statistical software, fill in the blanks below.

The regression equation is

Corn = _____ + _____ Year + _____ Soybeans

Why do the coefficients of year and soybeans differ from those in the previous two exercises? To answer this recall that, in general, for the multiple linear regression model

$$y_i = \beta_0 + \beta_1 x_{i1} + \beta_2 x_{i2} + \ldots + \beta_p x_{ip} + \varepsilon_i$$

the estimate of β_j represents the effect of the explanatory variable x_j when it is added to a model already containing the other explanatory variables.

d) Summarize the results of the significance tests for the regression coefficients for year and soybean yield by completing the table below. You should be able to find the necessary information from your statistical software.

Variable	Parameter Estimate	Std. Error	t-ratio	P-value
Constant				
Year				
Soybeans				

What do you conclude?

e) Recall that a 95% confidence interval for β_j is

$$b_j \pm t^* \mathrm{SE}_{b_j}$$

where t^* is the upper 0.025 critical value for the $t(n - p - 1) = t(37)$ distribution, which in this case (see Table D in the text) is approximately 2.03. SE_{b_1} is the standard error of b_j and can be found as an entry in the table you completed in (d). Thus here our 95% confidence intervals are

For year: $b_1 \pm 2.03\,\mathrm{SE}_{b_1} =$

For soybeans: $b_2 \pm 2.03\,\mathrm{SE}_{b_2} =$

f) Use statistical software to make the residual plots. Do you see any patterns? (How do these plots compare to those in the previous two exercises?) What do you conclude?

COMPLETE SOLUTIONS

Exercise 11.44

a) A summary of the results of a simple linear regression analysis are given below. The numbers were obtained from Minitab but have been reformatted.

Source	Degrees of freedom	Sum of Squares	Mean Square	F
Model	1	18650	18650	192.04
Error	38	3690	97	
Total	39	22340		

The regression equation is

$$\text{Corn} = -3606 + 1.87\,\text{Year}$$

Variable	Parameter Estimate	Std. Error	t-ratio	P-value
Constant	-3605.6	266.8	-13.51	0.000
Year	1.8706	0.1350	13.86	0.000

$$s = \sqrt{\mathrm{MSE}} = 9.855 \quad R^2 = 83.5\%$$

From these results, we see that the *P*-value for the significance test for the slope is very small (below 0.001). This can be seen in the entry for the *P*-value of the parameter estimate for Year (which is the slope). Equivalently, one can use the ANOVA *F* statistic too and Table E to determine this *P*-value.

b) A normal quantile plot (from Minitab) of the residuals is given below.

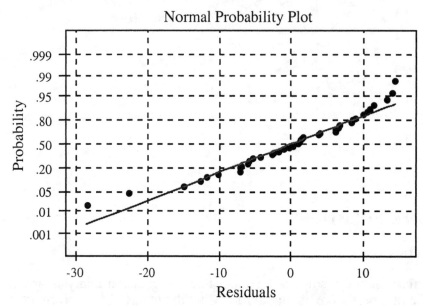

These points appear to lie reasonably close to a straight line and there are no striking features which would suggest that the data are not normal.

c) A plot of the residuals versus soybean yield is given below.

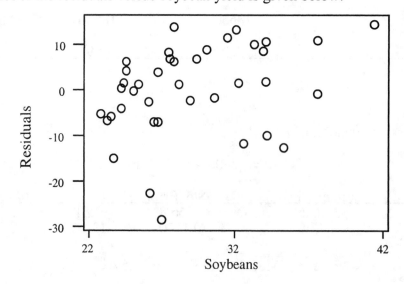

There is some evidence to an upward trend in the plot, suggesting that soybean yield might be useful in a multiple linear regression model with year to predict corn yield.

Exercise 11.45

a) A summary of the results of a simple linear regression analysis are given below. The numbers were obtained from Minitab but have been reformatted.

Source	Degrees of freedom	Sum of Squares	Mean Square	F
Model	1	19570	19570	268.44
Error	38	2770	73	
Total	39	22340		

The regression equation is

$$\text{Corn} = -47.2 + 4.80 \text{ Soybeans}$$

Variable	Parameter Estimate	Std. Error	t-ratio	P-value
Constant	-47.199	8.578	-5.50	0.000
Soybeans	4.8031	0.2932	16.38	0.000

$$s = \sqrt{\text{MSE}} = 8.538 \quad R^2 = 87.6\%$$

From these results, we see that the P-value for the significance test for the slope is very small (below 0.001). This can be seen in the entry for the P-value of the parameter estimate for Soybeans (which is the slope). Equivalently, one can use the ANOVA F statistic too and Table E to determine this P-value.

b) A normal quantile plot (from Minitab) of the residuals is given below.

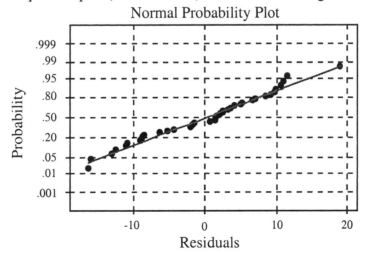

Normal Probability Plot

These points appear to lie reasonably close to a straight line and there are no striking features which would suggest that the data are not normal.

c) A plot of the residuals versus year is given below.

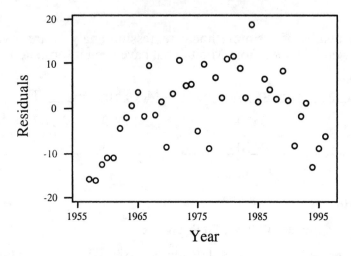

This plot has a curved trend, rising until around 1980 then falling. This suggests that the square of year (rather than year) might be useful in a multiple linear regression model with soybean yield to predict corn yield.

Exercise 11.46

a) The ANOVA table is

Source	Degrees of freedom	Sum of Squares	Mean Square	F
Model	2	20446	10223	199.73
Error	37	1894	51	
Total	39	22340		

$$s = \sqrt{\text{MSE}} = 7.154 \quad R^2 = 91.5\%$$

The null and alternative hypotheses tested by the ANOVA F test are

$$H_0: \beta_1 = \beta_2 = 0$$

$$H_a: \text{at least one of } \beta_1 \text{ and } \beta_2 \text{ is not } 0$$

where β_1 is the regression coefficient of year and β_2 the regression coefficient of soybean yield in our multiple linear regression model. The ANOVA F statistic has an $F(2, n - 3) = F(2, 37)$ distribution. If your statistical software does not give the P-value, you can approximate it from Table E in which case one finds the P-value is less than 0.001. We conclude that at least one of the two regression coefficients is different from 0 in the population regression equation.

b) In our multiple regression model, from our statistical software we find (see (a)) $R^2 = 91.5\%$. This is a little higher than for the simple linear regression model with year as the explanatory variable, where we found $R^2 = 83.5\%$. It is also slightly higher than for the simple linear regression model with soybean yield as the explanatory variable where we found $R^2 = 87.6\%$. Thus the multiple linear regression model explains a little bit higher percentage of the variability in the corn yield than either of the simple linear regression models of the previous two exercises.

c) From our statistical software (Minitab in our case) we get the fitted model

$$\text{Corn} = -1638 + 0.831 \text{ Year} + 2.98 \text{ Soybeans}$$

The coefficients differ because they have a different meaning here than in the simple linear regression models. The regression coefficient for year tells us the effect of year as an explanatory variable for corn yield in a model that includes the explanatory variable soybean yield. Soybean yield may capture some of the effect of year, producing a different value for the regression coefficient than we found in the simple linear regression model. Soybean yield was not present as an explanatory variable in the simple linear regression with year as an explanatory variable. Likewise, the regression coefficient for soybean yield tells us the effect of soybean yield as an explanatory variable for corn yield in a model that includes the explanatory variable year. Year may capture some of the effect of soybean yield, producing a different value for the regression coefficient than we found in the simple linear regression model. Year was not present as an explanatory variable in the simple linear regression with soybean yield as an explanatory variable.

d) From our statistical software (Minitab in our case) we get the significance test results for the regression coefficients for year and soybean yield to be

Variable	Parameter Estimate	Std. Error	t-ratio	P-value
Constant	-1637.9	384.5	-4.26	0.000
Year	0.8314	0.2009	4.14	0.000
Soybeans	2.9837	0.5036	5.92	0.000

We see that there is strong (P-values below 0.001) statistical evidence that both regression coefficients differ from 0 in the multiple linear regression model.

e) Our 95% confidence intervals are

For year: $b_1 \pm 2.03\, SE_{b_1} = 0.8314 \pm 2.03(0.2009) = 0.8314 \pm 0.4078$

For soybeans: $b_2 \pm 2.03\, SE_{b_2} = 2.9837 \pm 2.03(0.5036) = 2.9837 \pm 1.0223$

f) Plots of the residuals versus year and versus soybean yield are given below.

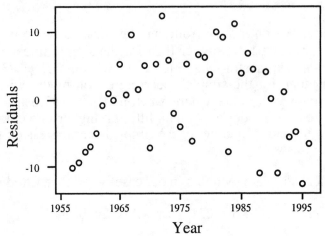

This plot has a curved trend, rising until around 1980 then falling. This suggests that the square of year might be useful in a multiple linear regression model with soybean yield and year to predict corn yield.

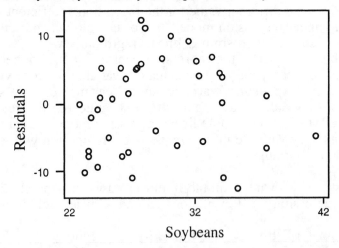

This plot shows no obvious trend, although there may be the slightest hint of a slightly curved pattern (residuals higher in the middle than at the ends). This may be due to the pattern observed in the residual plot versus year and the fact that year and soybeans are probably correlated.

CHAPTER 12

ONE-WAY ANALYSIS OF VARIANCE

OVERVIEW

The **one-way analysis of variance** is a generalization of the two sample t procedures. It allows comparison of more than two populations based on independent SRSs from each. As in the pooled two-sample t procedures, the populations are assumed to be normal with possibly different means, but with a common standard deviation. In the one-way analysis of variance we are interested in making formal inferences about the population means.

The simplest graphical procedure to compare the populations is to give side-by-side boxplots (see Chapter 1). Normal quantile plots can be used to check for extreme deviations from normality or outliers. The summary statistics required for the analysis of variance calculations are the means and standard deviations of each of the samples. An informal procedure to check the assumption of equal variances is to make sure that the ratio of the largest to the smallest standard deviation is less than 2. If the standard deviations satisfy this criterion and the normal quantile plots seem satisfactory, then the one-way ANOVA is an appropriate analysis.

The **F statistic** computed in the ANOVA table can be used to test the **null hypothesis** that the population means are all equal. The **alternative hypothesis** is that at least two of the population means are not equal. Rejection of the null hypothesis does not provide any information as to which of the population means are different.

If the researcher has specific questions about the population means before examining the data, these questions can often be expressed in terms of **contrasts**. Tests and confidence intervals about contrasts provide answers to these questions and allow the researcher to say more about which population means are different and what the sizes of these differences are.

When there are no specific questions before examining the data, **multiple comparisons** are often used to follow up rejection of the null hypothesis in a one-way analysis of variance. These multiple comparisons are designed to determine which pairs of population means are different and to give confidence intervals for the differences.

SAMPLE PROBLEMS

GUIDED SOLUTIONS

Exercise 12.33

KEY CONCEPTS - relationships among entries in the ANOVA table

a) The key relationships among the entries in the ANOVA table required to complete the table are

- the mean squares are equal to the sums of squares divided by their degrees of freedom

- the sum of squares for groups and the sum of squares for error add to the total sum of squares

- the degrees of freedom for groups and the degrees of freedom for error add to the total degrees of freedom (which is $N - 1$)

- the F statistic is the mean square for groups divided by the mean square for error

Use these relationships to complete the ANOVA table below.

Source	Degrees of freedom	Sum of squares	Mean square	F
Groups	3		158.96	
Error	32		62.81	
Total				

b) In the one-way ANOVA table, the F statistic is always used to test the null hypothesis that the means are equal and the alternative hypothesis that they are not equal. There aren't one- and two-sided tests as there were in some of the previous chapters. When writing down these hypotheses, you should try and phrase them in language or notation related to the particular data set.

c) Under the null hypothesis the F statistic has an $F(I - 1, N - I)$ distribution. The degrees of freedom for the F always agree with the degrees of freedom for groups and error, respectively in the ANOVA table. This exercise can be done by identifying the values of I and N in this data set, or by reading the appropriate degrees of freedom from the ANOVA table. You then need to go to Table E and see which F critical values in the table correspond to the value of the F statistic computed in this exercise.

d) Which entry in the ANOVA table corresponds to the estimate of the within-group variance?

Exercise 12.41

KEY CONCEPTS - comparisons among the means in the one-way ANOVA, contrasts

We reproduce the data for the four groups below.

Group	n	$\bar{x}$	s
Treatment (T)	10	51.90	6.42
Control (C)	5	57.40	10.46
Joggers (J)	11	49.73	6.27
Sedentary (S)	10	58.20	9.49

a) Contrasts are comparisons among the means designed to answer specific questions. These questions are posed before the data are collected. There are four means in this problem, μ_T, μ_C, μ_J and μ_S. Give each of the contrasts as a combination of some of these means. Remember that the sum of the coefficients of the means will add to zero for a contrast. The specific question being asked can also be stated in terms of hypotheses about the contrast.

1) Is T better than C? The contrast is $\psi_1 = \mu_C - \mu_T$. The hypotheses are $H_0: \psi_1 = 0$ and $H_a: \psi_1 > 0$. As usual, the alternative is the effect we are trying to find and $\psi_1 > 0$ corresponds to $\mu_C > \mu_T$ or T better than C, since lower depression scores are "better."

2) Is T better than the average of C and S? The contrast is $\psi_2 =$

The hypotheses are H_0:   and H_a:

      3) Is J better than the average of the other three groups? The contrast is
$\psi_3 =$

The hypotheses are H_0:   and H_a:

b) The estimate of a contrast or combination of population means $\psi = \sum a_i \mu_i$ is given by the same combination of the sample means. The sample contrast or estimator of ψ is $c = \sum a_i \bar{x}_i$. The standard error of the contrast is given by

$$SE_c = s_p \sqrt{\sum \frac{a_i^2}{n_i}}$$

and a test of the null hypothesis that the contrast is 0 uses the t statistic, $t = \dfrac{c}{SE_c}$. The first thing that needs to be done when applying these results in a particular setting is to identify the a_i in the contrast. In the three contrasts from (a) of this exercise, the first is a simple difference of means, while the second and third are more complicated. The answer below applies these general results about contrasts to the second contrast, leaving the the first and third contrast for you to practice on.

      1) Is T better than C?

Sample contrast =

Standard error =

t statistic =   df = P-value =

      2) Is T better than the average of C and S? The contrast is

$$\psi_2 = \frac{1}{2}(\mu_C + \mu_S) - \mu_T$$

Remember the sample contrast is the same combination as the population contrast with the population means replaced by the sample means.

$$\text{Sample contrast} = \frac{1}{2}(\bar{x}_C + \bar{x}_S) - \bar{x}_T = \frac{1}{2}(57.40 + 58.20) - 51.90 = 5.9$$

The standard error of a contrast requires the within-groups estimate of the standard deviation s_p, and the a_i and n_i for the contrast. The within-groups estimate of the standard deviation s_p, corresponds to the square root of the MSE in the ANOVA table. The ANOVA table for this data is given in Exercise 12.23 and $s_p = \sqrt{62.81} = 7.93$.

The a_i for this contrast are 1/2, 1/2, and -1. These a_i correspond to sample sizes n_i of 5, 10, and 10 respectively. We can now apply the general formula for the standard error of a contrast.

$$\text{Standard error} = \text{SE}_c = s_p\sqrt{\sum\frac{a_i^2}{n_i}} = 7.93\sqrt{\frac{(1/2)^2}{5}+\frac{(1/2)^2}{10}+\frac{(-1)^2}{10}} = 3.3174$$

$$t \text{ statistic} = \frac{c}{\text{SE}_c} = \frac{5.9}{3.3174} = 1.78$$

Since the alternative is one-sided, the P-value is $P(t > 1.78)$ where the t has the degrees of freedom that are associated with s_p which is $N - I = 32$. Using Table D we see that the P-value is between 0.025 and 0.05, or using statistical software the value is 0.0423.

   3) Is J better than the average of the other three groups?

Sample contrast =

Standard error =

t statistic =   df =   P-value =

Does the use of contrasts give an adequate summary of the results? How does the information from the contrasts compare with the information contained in simply presenting the ANOVA table and the associated F statistic?

c) Is this observational data or an experiment? Can we conclude that there is a cause and effect relationship or is there only an association?

Exercise 12.45

KEY CONCEPTS - Bonferroni multiple comparisons

Multiple comparison methods are designed to determine which pairs of means are different. These methods are used when the null hypothesis is rejected, and we didn't have specific questions about the means in advance of the analysis. Multiple comparison procedures are performed by computing t statistics for all pairs of means using the fomula

$$t_{ij} = \frac{\bar{x}_i - \bar{x}_j}{s_p\sqrt{\frac{1}{n_i} + \frac{1}{n_j}}}$$

This is the same t statistic that would be used when studying the contrast $\psi = \mu_i - \mu_j$. We declare the population means μ_i and μ_j different whenever

$$|t_{ij}| \geq t^{**}$$

where the value of t^{**} depends on the multiple comparison procedure being applied. In this exercise, the Bonferroni multiple comparison procedure is being used and the value of t^{**} is given in the exercise as $t^{**} = 2.53$. The simulaneous confidence intervals which tell you about the the *sizes* of the differences between the means are found using the formula

$$\bar{x}_i - \bar{x}_j \pm t^{**} s_p\sqrt{\frac{1}{n_i} + \frac{1}{n_j}}$$

When doing the calculations, it is often simplest to summarize them in a table like the one on the next page. The details for the first pair of means (T, C) are worked out for you. Note that the quantity $s_p\sqrt{\frac{1}{n_i} + \frac{1}{n_j}}$ which appears in the calculations of both the t statistic and the confidence interval, has the same value for any two groups with the same pair of sample sizes such as (T, C) and (C, S). Use of this fact will save some time in the calculations.

For the pair (T, C),

$$\bar{x}_T - \bar{x}_C = 51.90 - 57.40 = -5.5$$

$$s_p\sqrt{\frac{1}{n_T} + \frac{1}{n_C}} = 7.93\sqrt{\frac{1}{10} + \frac{1}{5}} = 4.3434$$

The t statistic is

$$t_{TC} = \frac{\bar{x}_T - \bar{x}_C}{s_p\sqrt{\dfrac{1}{n_T} + \dfrac{1}{n_C}}} = \frac{-5.5}{4.3434} = -1.27$$

The confidence interval for $\mu_T - \mu_C$ is

$$\bar{x}_T - \bar{x}_C \pm t^{**} s_p\sqrt{\frac{1}{n_T} + \frac{1}{n_C}} = -5.5 \pm 2.53 \times 4.3434 = -5.5 \pm 10.99$$

Pair of means $\bar{x}_i - \bar{x}_j$		$s_p\sqrt{\dfrac{1}{n_i} + \dfrac{1}{n_j}}$	t_{ij}	Confidence interval
(T, C)	-5.50	4.3434	-1.27	-17.06 ± 10.99
(T, J)				
(T, S)				
(C, J)				
(C, S)				
(J, S)				

How would you interpret these results?

COMPLETE SOLUTIONS

Exercise 12.31

a) The mean square for groups is the group sum of squares divided by its degrees of freedom. We are given the mean square for groups, so the group sum of squares is the mean square multiplied by its degrees of freedom, 3 x 158.96 = 476.88. Similarly, the sum of squares for error is 32 x 62.81 = 2009.92. The other entries follow the rules given in the guided solution.

Source	Degrees of freedom	Sum of squares	Mean square	F
Groups	3	476.88	158.96	2.53
Error	32	2009.92	62.81	
Total	35	2486.80		

b) The null hypothesis is that the mean depression scores for the four populations from which the samples are assumed to be drawn are equal. The alternative hypothesis is that the means are different, but does not specify what these differences might be. Using the notation, we have the hypotheses

$$H_0: \mu_T = \mu_C = \mu_J = \mu_S \text{ and } H_a: \mu_T, \mu_C, \mu_J, \text{ and } \mu_S \text{ are not all equal}$$

c) Under the null hypothesis the F statistic has an $F(I - 1, N - I) = F(3, 32)$ distribution since the number of groups $I = 4$ and the number of subjects $N = 36$. Referring to Table E, we find that the F critical values for tail probabilities 0.10 and 0.05 are 2.28 and 2.92 respectively, where we have gone to the $F(3, 30)$ distribution to be conservative, since 32 degrees of freedom is not in the table. The P-value is between 0.05 and 0.10. Computer software using the degrees of freedom 3 and 32 gives the P-value as 0.0747. There is some evidence of a difference in the depression scores between the four groups, but the evidence is not that strong.

d) The mean square for error is the estimate of the within-group variance, so $s_p^2 = 62.81$. The square root is $s_p = \sqrt{62.81} = 7.93$.

Exercise 12.41

a) The first contrast is given in the Guided Solutions.

2) Is T better than the average of C and S? The contrast is $\psi_2 = \frac{1}{2}(\mu_C + \mu_S) - \mu_T$.

The hypotheses are $H_0: \psi_2 = 0$ and $H_a: \psi_2 > 0$. The alternative consists of parameter values for which the average of C and S is greater than T, or since lower scores are better, that T is better than the average of C and S.

3) Is J better than the average of the other three groups? The contrast is $\psi_3 = \frac{1}{3}(\mu_C + \mu_S + \mu_T) - \mu_J$. The hypotheses are $H_0: \psi_3 = 0$ and $H_a: \psi_3 > 0$.

b)

1) Is T better than C? $\psi_1 = \mu_C - \mu_T$.

Sample contrast = $\bar{x}_C - \bar{x}_T = 57.40 - 51.90 = 5.5$

$$\text{Standard error} = \text{SE}_c = s_p\sqrt{\sum\frac{a_i^2}{n_i}} = 7.93\sqrt{\frac{1^2}{5} + \frac{(-1)^2}{10}} = 4.34.$$

$$t \text{ statistic} = \frac{c}{\text{SE}_c} = \frac{5.5}{4.34} = 1.27 \qquad df = 32, \; P\text{-value} = 0.1066 \text{ (using software)}$$

2) See guided solution.

3) The contrast is $\psi_3 = \dfrac{1}{3}(\mu_C + \mu_S + \mu_T) - \mu_J$.

$$\text{Sample contrast} \qquad = \frac{1}{3}(\bar{x}_C + \bar{x}_S + \bar{x}_T) - \bar{x}_J$$

$$= \frac{1}{3}(57.40 + 58.20 + 51.90) - 49.73 = 6.10$$

$$\text{Standard error} = \text{SE}_c = s_p\sqrt{\sum\frac{a_i^2}{n_i}} = 7.93\sqrt{\frac{(1/3)^2}{5} + \frac{(1/3)^2}{10} + \frac{(1/3)^2}{10} + \frac{(-1)^2}{11}}$$

$$= 2.9175.$$

$$t \text{ statistic} = \frac{c}{\text{SE}_c} = \frac{6.1}{2.9175} = 2.09 \quad df = 32, \; P\text{-value} = 0.0223 \text{ (using software)}$$

The use of contrasts provides specific answers as to the nature of some of the differences between the four groups. The ANOVA table and associated F statistic just tell us whether there is evidence of a difference between the four groups, but we can't say more. The use of contrasts allows us to identify where the differences are and answer specific questions about comparisons between the groups which are of interest.

c) This is observational data. The experimenter did not assign the male subjects to be in an exercise group, jog, or have a sedentary lifestyle. There is a possibility the groups differed on other (confounding) variables which might be related to depression. The description of the data in Exercise 12.32 indicates that the experimenter tried to select subjects similar in age and other characteristics. While this is desirable and may eliminate certain potential confounding variables such as age, we can't be sure that the experimenter has balanced the groups on all potential confounding variables. Randomization of subjects to the treatment groups is still the best way to do this.

Exercise 12.45

Below is the completed table for all pairwise comparisons using the Bonferroni method.

Pair of means	$\bar{x}_i - \bar{x}_j$	$s_p\sqrt{\dfrac{1}{n_i} + \dfrac{1}{n_j}}$	t_{ij}	Confidence interval
(T, C)	-5.50	4.3434	-1.27	-5.50 ± 10.99
(T, J)	2.17	3.4649	0.63	2.17 ± 8.77
(T, S)	-6.30	3.5464	-1.77	-6.30 ± 8.97
(C, J)	7.67	4.2771	1.79	7.67 ± 10.82
(C, S)	-0.80	4.3434	-0.18	-0.80 ± 10.99
(J, S)	-8.47	3.4649	-2.44	-8.47 ± 8.77

Using an error probability of 0.05 we would conclude that there is no evidence of a difference between any pair of means since none of these t_{ij} values exceeds 2.53 in absolute value. This is the same conclusion we reached when using the F statistic in the ANOVA table.

The interpretation of 95% *simultaneous* confidence intervals is that, in the long run, we are requiring that 95% of the time *all six* intervals contain the true differences in means, rather than only requiring this property to hold for a single interval. In order to accomplish this, the intervals need to be much longer than an individual confidence interval. Similarly, when doing the hypothesis tests, the error probability is controlling the probability of *any* false rejection among the six tests, so the t^{**} value is much larger than it would be for an individual test. The multiple comparison approach can be quite conservative and it is not surprising that some of the contrasts in Exercise 12.41 were significantly different from zero, while based on the multiple comparison procedure we would conclude all pairs of means are equal. To overcome the conservative nature of multiple comparisons, some researchers choose an error probability that is a little higher than you might ordinarily use when performing a single test, such as 0.10 for example.

Finally, remember that not rejecting the null hypothesis is not the same as demonstrating that it is true. It is still a good idea to look at the confidence intervals and the estimates of the differences, to see if any of the comparisons are deserving of further study, possibly with a larger sample size.

CHAPTER 13

TWO-WAY ANALYSIS OF VARIANCE

OVERVIEW

The two-way analysis of variance is designed to compare the means of populations that are classified according to two factors. As with the one-way ANOVA, the populations are assumed to be normal with possibly different means and the same standard deviation. The observations are independent SRSs drawn from each population.

The preliminary data summary should include examination of means and standard deviations, and normal quantile plots. Typically the means are summarized in a two-way table with the rows corresponding to the level of one factor and the columns corresponding to the levels of the second factor. The **marginal means** are computed by taking averages of these cell means across the rows and columns. These means are typically plotted so that the **main effects** of each factor can be examined as well as their **interaction**.

In the two-way ANOVA table, the **model** variation is broken down into parts due to each of the main effects and a third part due to the interaction. In addition, the ANOVA table organizes the calculations required to compute F statistics and P-values to test hypotheses about these main effects and their interaction. The within-group variance is estimated by pooling the standard deviations from the cell and corresponds to the mean square for error in the ANOVA table.

GUIDED SOLUTIONS

Exercise 13.9

KEY CONCEPTS - plotting group means, main effects, interactions

a) Complete the plot of the mean GITH for these diets on the axes provided below. You should plot the two means when Chromium is Low above the symbol L, and the two means when Chromium is Normal above the symbol N. Then, for each Eat group, connect the points for the two Chromium means.

Interaction Plot - Means for GITH

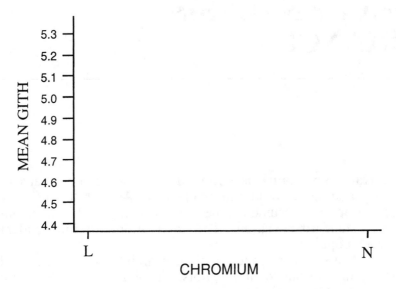

b) In Exercise 13.23 we will see which effects in the plot are real and which are due to chance variation. For now, try to identify some of the main features of the plot. Do the lines seem to be approximately parallel? Is the effect of Chromium similar for both levels of Eat? Are the lines for the two levels of Eat far apart? Is there much change in the mean GITH when going from Low to Normal levels of Chromium? Think about what the answers to these different questions are telling you about the factors Eat and Chromium.

c) In the table below fill in the marginal means.

Chromium	Eat	
	M	R
L	4.545	5.175
N	4.425	5.317

Compare the difference between the M and R diets for each level of Chromium. How does this comparison show up in your plot?

Exercise 13.13

KEY CONCEPTS - computing group means and marginal means, two-way ANOVA

a) In the table below fill in the sample sizes, means and standard deviations for the twelve material-time groups. You should try and do the computations using computer software.

group	N	Mean	StDev
ECM1,4			
ECM1,8			
ECM2,4			
ECM2,8			
ECM3,4			
ECM3,8			
MAT1,4			
MAT1,8			
MAT2,4			
MAT2,8			
MAT3,4			
MAT3,8			

In the table below fill in the means for the different size-species groups and the marginal means for size and species. This type of table is produced easily using most computer software packages.

```
ROWS: TIME      COLUMNS: MATERIAL

        ECM1    ECM2    ECM3    MAT1    MAT2    MAT3    ALL

4
8
ALL
```

b) Plot the means of the twelve groups computed in (a) on the axes given. The time is on the x axis and the mean % Gpi is on the y axis. For each material, connect the two means corresponding to the different times.

Interaction Plot - Means for % Gpi

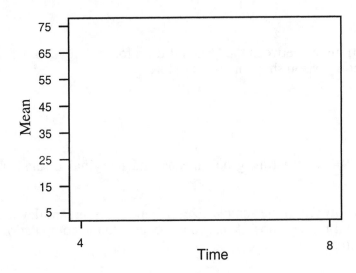

Describe any patterns you see. Focus on the differences between times, materials and any interaction between them.

c) Use your computer software to generate an ANOVA table similar to the one below.

```
Analysis of Variance for % Gpi
```

Source	DF	SS	MS	F	P
Material	5	27045.1	5409.0	251.26	0.000
Time	1	6.2	6.2	0.29	0.595
Material*Time	5	6.2	1.3	0.06	0.998
Error	24	516.7	21.5		
Total	35	27574.3			

Are the results of the significance tests in agreement with the impressions that you formed from the graphs in (b)? If the results are not what you expected, try to come up with a reason for the discrepancies between the ANOVA table and the plots.

Exercise 13.23

KEY CONCEPTS - two-way ANOVA table, drawing conclusions

a) Complete the ANOVA table below. The first step is to fill in the degrees of freedom. Then the mean squares can be obtained easily from the sums of squares and the F's as the ratio of the appropriate mean squares.

Source	Degrees of freedom	Sum of squares	Mean square	F
A(Chromium)			0.00121	
B(Eat)			5.79121	
AB			0.17161	
Error			0.03002	
Total				

b) The degrees of freedom for an F statistic are the degrees of freedom for the mean square in the numerator and the degrees of freedom for the mean square in the denominator, respectively. When you have determined the value of the F statistic and its degrees of freedom, you need to compare the F-value to the critical values provided in Table E. When using Table E the best that can be done is to give two values that the P-value must fall between. What is your conclusion?

c) Carry out the tests for the main effects of Chromium and Eat. What do you conclude?

d) The within-group variance corresponds to which entry in the ANOVA table?

e) Summarize the results of the experiment, using what you have learned in this exercise and in Exercise 13.9.

COMPLETE SOLUTIONS

Exercise 13.7

a)

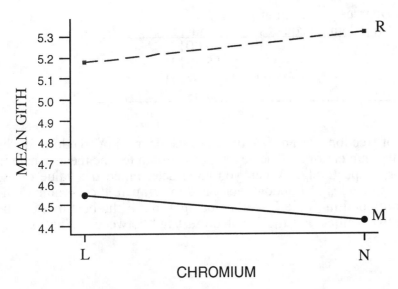

Interaction Plot - Means for GITH

b) The plot suggests that there may be an interaction. The effect of Chromium when going from Low to Normal levels is to decrease mean GITH when the rats could eat as much as they wanted (M) and to increase the mean GITH when the total amount the rats could eat was restricted (R). Without a formal hypothesis test, we don't know if this apparent interaction is due to chance variation, or whether the effect is real. In terms of the effect of Chromium, it appears to be small compared to the effect of Eat. The two lines are quite far apart showing a larger effect of Eat. The change in going from the Low to Normal levels of Chromium is much smaller.

c)

	Eat		Mean
Chromium	M	R	
L	4.545	5.175	4.860
N	4.425	5.317	4.871
Mean	4.485	5.246	4.866

At the Low level of Chromium the difference between M and R is -0.63, and at the Normal level of Chromium the difference between M and R is -0.892. In the plot the two means at the Low level of Chromium are closer together than the two means at the Normal level of Chromium. This is reflected in the fact that the two lines are not parallel.

Exercise 13.13

a) The table below gives the sample sizes, means and standard deviations for the twelve material-time groups.

Variable	group	N	Mean	StDev
% Gpi	ECM1,4	3	65.00	8.66
	ECM1,8	3	63.33	2.89
	ECM2,4	3	63.33	2.89
	ECM2,8	3	63.33	5.77
	ECM3,4	3	73.33	2.89
	ECM3,8	3	73.33	5.77
	MAT1,4	3	23.33	2.89
	MAT1,8	3	21.67	5.77
	MAT2,4	3	6.67	2.89
	MAT2,8	3	6.67	2.89
	MAT3,4	3	11.67	2.89
	MAT3,8	3	10.00	5.00

The table below gives the means for the different size-species groups and the marginal means for size and species.

ROWS: TIME   COLUMNS: MATERIAL

	ECM1	ECM2	ECM3	MAT1	MAT2	MAT3	ALL
4	65.000	63.333	73.333	23.333	6.667	11.667	40.556
8	63.333	63.333	73.333	21.667	6.667	10.000	39.722
ALL	64.167	63.333	73.333	22.500	6.667	10.833	40.139

b)

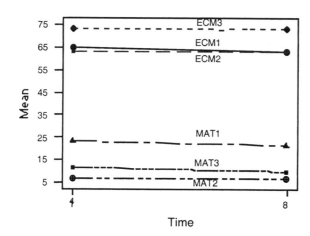

Interaction Plot - Means for % Gpi

The most striking feature in the plot is the complete lack of a time effect. The mean % Gpi at 4 and 8 weeks are almost identical and as a result there is almost no interaction. The important effect is the material, with the ECM (extracellular) material having a much higher % Gpi than the MAT (inert) material.

c) The ANOVA table is reproduced below.

```
Analysis of Variance for % Gpi

Source          DF        SS        MS        F        P
Material         5   27045.1    5409.0   251.26   0.000
Time             1       6.2       6.2     0.29   0.595
Material*Time    5       6.2       1.3     0.06   0.998
Error           24     516.7      21.5
Total           35   27574.3
```

The ANOVA table supports the interaction plot in (b). There is no evidence of a time effect or an interaction between time and material. However there is a highly significant effect of material. Note that the ANOVA table does not tell which materials are different or whether there is a difference between the ECM (extracellular) material and the MAT (inert) material. These types of conclusions require further multiple comparisons or examination of specific contrasts as in the one-way analysis of variance.

Exercise 13.23

a) The degrees of freedom for the main effects are the number of levels of the factor minus 1. The degrees of freedom for Chromium are 2 - 1 = 1 and for Eat are 2 - 1 = 1. The degrees of freedom for the interaction is the product of the degrees of freedom for the main effects and is $1 \times 1 = 1$. The total degrees of freedom is the number of subjects minus 1 which is 40 - 1 = 39. The degrees of freedom for error can be gotten by subtraction. The mean squares are the sums of squares divided by their degrees of freedom, and the F for main effects and interaction are gotten by dividing the mean squares for each by the error mean square.

Source	Degrees of freedom	Sum of squares	Mean square	F
A(Chromium)	1	0.00121	0.00121	0.04
B(Eat)	1	5.79121	5.79121	192.89
AB	1	0.17161	0.17161	5.72
Error	36	1.08084	0.03002	
Total	39	7.04487		

b) Under the null hypothesis, the F statistic used to test the null hypothesis of no interaction has an $F(1, 36)$ distribution. The numerical value of the F statistic is 5.72. Referring this to the critical values in Table E, we see the P-value is between 0.025 and 0.010. Since 36 degrees of freedom is not in the table, we need to look at the entries for 30 and 40. Computer software using the $F(1, 36)$ distribution gives the P-value as 0.0221.

c) The main effect of Chromium has an $F = 0.04$. From Table E, the P-value is greater than 0.10. Computer software using the $F(1, 36)$ distribution gives the P-value as 0.8426. The main effect of Eat has an $F = 192.89$. From Table E, the P-value is less than 0.001. Computer software using the $F(1, 36)$ distribution gives the P-value as 0.0000.

d) The within-group variance is the mean square for error, so $s_p^2 = 0.03002$, and $s_p = \sqrt{0.03002} = 0.173$.

e) The interaction between Eat and Chromium is statistically significant, but the effect is relatively small compared to the main effect of Eat. It would be up to the experimenter to determine whether the difference in the effect of going from Low to Normal levels of Chromium for the two levels of Eat is large enough to be of practical interest. By far, the biggest effect is the main effect of Eat. The mean GITH scores are much smaller when the rats could eat as much as they wanted (M) than when the total amount the rats could eat was restricted (R).

CHAPTER 14

NONPARAMETRIC TESTS

SECTION 14.1

OVERVIEW

Many of the statistical procedures described in previous chapters assumed that the samples were drawn from normal populations. **Nonparametric tests** do not require any specific form for the distributions of the populations from which the samples were drawn. Many nonparametric tests are **rank tests**; that is, they are based on the **ranks** of the observations rather than on the observations themselves. When ranking the observations from smallest to largest, tied observations receive the average of their ranks.

The **Wilcoxon rank sum test** compares two distributions. The objective is to determine if one distribution has systematically larger values than the other. The observations are ranked, and the **Wilcoxon rank sum statistic W** is the sum of the ranks of one of the samples. The Wilcoxon rank sum test can be used in place of the **two-sample t test** when samples are small or the populations are far from normal.

Exact P-values require special tables and are produced by some statistical software. However, many statistical software packages give only approximate P-values based on a normal approximation, typically with a continuity correction employed. Many packages also make an adjustment in the normal approximation when there are ties in the ranks.

GUIDED SOLUTIONS

Exercise 14.3

KEY CONCEPTS - ranking data, two-sample problem, Wilcoxon rank sum test

a) Order the observations from smallest to largest in the space provided. Use a different color or underline those observations in the high progress group. This will make it easier to determine the ranks assigned to each group.

b) Now suppose the first sample is the high progress group and the second sample is the low progress group. The choice of which sample we call the first sample and which we call the second sample is arbitrary. However, the Wilcoxon rank sum test is the sum of the ranks of the first sample, and the formulas for the mean and variance of W distinguish between the sample sizes for the first and the second samples. What are the ranks of the high progress observations? Use these ranks to compute the value of W.

$W =$

What are the values of n_1, n_2 and N? Use these to evaluate the mean and standard deviation of W according to the formulas below.

$$\mu_W = \frac{n_1(N+1)}{2} =$$

$$\sigma_W = \sqrt{\frac{n_1 n_2 (N+1)}{12}} =$$

c) What kind of values would W have if the alternative were true? Use the normal approximation (with the continuity correction) to find the approximate P-value. If you have access to software or tables to evaluate the exact P-value, compare it with the approximation.

$$z = \frac{W - \mu_W}{\sigma_W} =$$

P-value $=$

What are your conclusions?

d) Order the observations from smallest to largest in the space provided. Use a different color or underline those observations in the high progress group. This will make it easier to determine the ranks assigned to each group. Remember ties are broken by assigning all tied values the average of the ranks they occupy.

COMPLETE SOLUTIONS

Exercise 14.3

a) The observations are first ordered from smallest to largest. The observations in bold are from the high progress group.

$$0.28, 0.38, 0.49, \mathbf{0.54}, 0.66, 0.77, \mathbf{0.79}, \mathbf{0.80}, \mathbf{0.82}, \mathbf{0.89}$$

b) Suppose the first sample is the high progress readers and the second sample is the low progress group. In this case, $n_1 = n_2 = 5$, and N, the sum of the sample sizes is 10. The high progress observations in bold have ranks 4, 7, 8, 9 and 10, and the sum of these ranks is $W = 38$. The values for the mean and variance are

$$\mu_W = \frac{n_1(N+1)}{2} = \frac{5(10+1)}{2} = 27.5 \text{ and}$$

$$\sigma_W = \sqrt{\frac{n_1 n_2 (N+1)}{12}} = \sqrt{\frac{(5)(5)(10+1)}{12}} = 4.787$$

c) We expect W to be large when the alternative hypothesis is true, as the high progress group should receive the larger ranks. If we use the continuity correction, we act as if the value of $W = 38$ occupies the interval from 37.5 to 38.5. This means to compute the P-value, we first calculate the probability that $W \geq 37.5$, giving

$$P(W \geq 37.5) = P\left(\frac{W - \mu_W}{\sigma_W} \geq \frac{37.5 - 27.5}{4.787}\right) = P(Z \geq 2.09) = 0.0183$$

The exact value using statistical software is 0.016, so the agreement is quite good when using the normal approximation with the continuity correction. There is strong evidence that the high progess readers tend to have higher scores than the low progress readers.

d) The observations are first ordered from smallest to largest. The observations in bold are from the high progress group.

0.00, 0.36, 0.40, 0.55, **0.55, 0.57, 0.70, 0.72,** 0.72, **0.84**

The ranks assigned to these observations are

1, 2, 3, 4.5, **4.5, 6, 7, 8.5,** 8.5, **10**

where ties have been broken using average ranks.

SECTION 14.2

OVERVIEW

The **Wilcoxon signed rank test** is a nonparametric test for matched pairs. It tests the null hypothesis that there is no systematic difference between the observations within a pair against the alternative that one observation tends to be larger.

The test is based on the **Wilcoxon signed rank statistic W^+**, which provides another example of a nonparametric test using ranks. The absolute values of the observations are ranked and the sum of the ranks of the positive (or negative) differences gives the value of W^+. The **matched pairs t test** and the **sign test** are two other alternative tests for this setting.

P-values can be found from special tables of the distribution or a normal approximation to the distribution of W^+. Some software computes the exact P-value and other software uses the normal approximation, typically with a ties correction. Many packages make an adjustment in the normal approximation when there are ties in the ranks.

GUIDED SOLUTIONS

Exercise 14.15

KEY CONCEPTS - matched pairs, Wilcoxon signed rank statistic

First give the null and alternative hypotheses

H_0:
H_a:

To compute the Wilcoxon signed rank statistic, first order the absolute values of the differences and rank them. Due to the large number of ties in this exercise, you need to be careful when computing the ranks. For any tied group of observations, they should each receive the average rank for the group. (Note that the positive observations are in bold.) In addition, you need to drop the observations that are equal to zero (the pretest and the posttest scores were the same) and reduce the sample size accordingly. What is the new value of n?

Absolute values	Ranks
1	1.5
1	1.5
2	4.0
2	4.0
2	4.0
3	8.5
3	8.5
3	8.5
3	8.5
3	8.5
3	8.5
-6	14.5
6	14.5
6	14.5
6	14.5
6	14.5
6	14.5

To see how the ranks are computed, the 1's would get ranks 1 and 2 so their average rank is 1.5. The 2's would get ranks 3, 4, and 5, so their average rank is 4.0, and so on. If W^+ is the sum of the ranks of the positive observations compute the value of W^+.

$W^+ =$

Evaluate the mean and standard deviation of W^+ according to the formulas below.

$$\mu_{W^+} = \frac{n(n+1)}{4}$$

$$\sigma_{W^+} = \sqrt{\frac{n(n+1)(2n+1)}{24}}$$

Now use the mean and standard deviation to compute the standardized rank sum statistic

$$z = \frac{W^+ - \mu_{W^+}}{\sigma_{W^+}} =$$

Do you expect W^+ to be small or large if the alternative is true? Use the normal approximation to find the approximate P-value. (Note: The continuity correction is often used for statistics whose values are whole numbers. This is true for W^+ when there are no tied observations. However, when there are tied observations and W^+ can take on values which are not whole numbers the continuity correction would not be employed.)

What are your conclusions?

COMPLETE SOLUTIONS

Exercise 14.15

The null and alternative hypotheses are

H_0: scores have the same distribution for the pre and posttest
H_a: posttest scores are systematically higher than pre-test scores

The Wilcoxon signed rank statistic is

$$W^+ = 1.5 + 1.5 + 4.0 + 4.0 + 4.0 + 8.5 + 8.5 + 8.5 + 8.5 + 8.5 + 8.5$$
$$+ 14.5 + 14.5 + 14.5 + 14.5 + 14.5 = 138.5$$

The values for the mean and variance are (recall that three observations had a value of zero and were dropped reducing n to 17)

$$\mu_{w^+} = \frac{n(n+1)}{4} = \frac{17(17+1)}{4} = 76.5 \text{ and}$$
$$\sigma_{w^+} = \sqrt{\frac{n(n+1)(2n+1)}{24}} = \sqrt{\frac{(17)(18)(34+1)}{24}} = 21.125$$

and the standardized signed rank statistic W^+ is

$$\frac{W^{+} - \mu_{w^{+}}}{\sigma_{w^{+}}} = \frac{138.5 - 76.5}{21.125} = 2.93$$

If posttest scores are systematically higher, we would expect the differences (posttest - pretest) to be positive. Thus the ranks of the positive observations should be large and we would expect the value of the statistic W^{+} to be large when the alternative hypothesis is true. The approximate P-value is $P(Z \geq 2.93) = 0.0017$.

The output from the MINITAB computer package gives a similar result. Many computer packages, including MINITAB, include a correction to the standard deviation in the normal approximation to account for the ties in the ranks.

```
Wilcoxon Signed Rank Test

TEST OF MEDIAN = 0.000000 VERSUS MEDIAN G.T.  0.000000

                  N FOR    WILCOXON
            N     TEST    STATISTIC   P-VALUE
CHANGE      17     17       138.5      0.002
```

The data shows that the scores on understanding of spoken French have improved after attending a summer language institute.

SECTION 14.3

OVERVIEW

The **Kruskal-Wallis test** is the nonparametric test for the **one-way analysis of variance** setting. In comparing several populations, it tests the null hypothesis that the distribution of the response variable is the same in all groups and the alternative hypothesis that some groups have distributions of the response variable that are systematically larger than others.

The **Kruskal-Wallis statistic** H compares the average ranks received for the different samples. If the alternative is true, some of these should be larger than others. Computationally, it essentially arises from performing the usual one-way ANOVA to the ranks of the observations rather than the observations themselves.

P-values can be found from special tables of the distribution or a chi-square approximation to the distribution of H. When the sample sizes are not too small, the distribution of H for comparing I populations has approximately a chi-square distribution with $I - 1$ degrees of freedom. Some software computes the exact P-value and other software uses the chi-square approximation, typically with an adjustment in the chi-square approximation when there are ties in the ranks.

GUIDED SOLUTIONS

Exercise 14.25

KEY CONCEPTS - one-way ANOVA, Kruskal-Wallis statistic

a) The Kruskal-Wallis test is testing

H_0: The distribution of insects trapped is the same for all colors
H_a: Some colors have systematically higher numbers of trapped insects

What are the values of I, the n_i and N in this example?

b) To compute the Kruskal-Wallis test statistic, the 20 observations are first arranged in increasing order. That step has been carried out below, where we have kept track of the group for each observation. You need to fill in the ranks in the line provided. Remember to use average ranks for tied observations.

Trapped	7	11	12	13	14	14	15
Group	Blue	Blue	White	White	White	Blue	Green
Rank							

Trapped	16	17	17	20	21	21	25
Group	Blue	White	White	Blue	Blue	White	Green
Rank							

Trapped	32	37	38	39	41	45	46
Group	Green	Green	Lemon	Green	Green	Lemon	Lemon
Rank							

Trapped	47	48	59
Group	Lemon	Lemon	Lemon
Rank			

Now fill in the table on the next page which gives the ranks for each of the colors, and the sum of ranks for each group.

Color	Ranks	Sum of Ranks
Lemon		
White		
Green		
Blue		

Use the sum of ranks for the four groups and the numerical values of the n_i and N obtained in part (a) to evaluate the Kruskal-Wallis statistic below.

$$H = \frac{12}{N(N+1)} \sum \frac{R_i^2}{n_i} - 3(N+1) \ =$$

The value of H is compared with critical values in Table G for a chi-square distribution with $I - 1$ degrees of freedom, where I is the number of groups. What is the P-value and what do you conclude?

COMPLETE SOLUTIONS

Exercise 14.25

a) $I = 4$, the n_i are each 6 and $N = 24$.

b) The computations required for the Kruskal-Wallis test statistic are summarized in the tables below.

Trapped	7	11	12	13	14	14	15
Group	Blue	Blue	White	White	White	Blue	Green
Rank	1	2	3	4	5.5	5.5	7

Trapped	16	17	17	20	21	21	25
Group	Blue	White	White	Blue	Blue	White	Green
Rank	8	9.5	9.5	11	12.5	12.5	14

Trapped	32	37	38	39	41	45	46
Group	Green	Green	Lemon	Green	Green	Lemon	Lemon
Rank	15	16	17	18	19	20	21

Trapped	47	48	59
Group	Lemon	Lemon	Lemon
Rank	22	23	24

Color	Ranks	Sum of Ranks
Lemon	17, 20, 21, 22, 23, 24	127
White	3, 4, 5.5, 9.5, 9.5, 12.5	44
Green	7, 14, 15, 16, 18, 19	89
Blue	1, 2, 5.5, 8, 11, 12.5	40

$$H = \frac{12}{N(N+1)} \sum \frac{R_i^2}{n_i} - 3(N+1)$$

$$= \frac{12}{24(24+1)} \left(\frac{127^2}{6} + \frac{44^2}{6} + \frac{89^2}{6} + \frac{40^2}{6} \right) - 3(24+1) = 16.98$$

Since $I = 4$ groups, the sampling distribution of H is approximately chi-square with $4 - 1 = 3$ degrees of freedom. From Table F we see the P-value is approximately 0.001. There is strong evidence of a difference in the number of insects trapped between the four groups. As in the one-way ANOVA, nonparametric multiple comparisons or contrasts would be required to explore the group differences further.

The MINITAB software gives the output below when doing the Kruskal-Wallis test. The medians, average ranks (in place of sums of ranks), H statistic and P-value are given. The H statistic with an adjustment for ties in the ranks is also given.

Kruskal-Wallis Test

```
LEVEL       NOBS     MEDIAN   AVE. RANK
    1          6      46.50       21.2
    2          6      15.50        7.3
    3          6      34.50       14.8
    4          6      15.00        6.7
OVERALL       24                  12.5

H = 16.95  d.f. = 3  p = 0.001
H = 16.98  d.f. = 3  p = 0.001 (adjusted for ties)
```

CHAPTER 15

LOGISTIC REGRESSION

OVERVIEW

In Chapter 8 we studied random variables y which can take on only two values (yes or no, success or failure, live or die, acceptable or not). It is usually convenient to code the two values as 0 (failure) and 1 (success) and let p denote the probability of a 1 (success). If y is observed on n independent trials, the total number of 1's can often be modeled as binomial with n trials and probability of success p. In this chapter, we consider methods that allow us to investigate how y depends on one or more explanatory variables. The methods of simple linear and multiple linear regression do not directly apply since the distribution of y is binomial rather than normal. However, it is possible to use a method, called **logistic regression**, which is similar to simple and multiple linear regression.

We define the **odds** as $p/(1-p)$, the ratio of the probability that the event happens to the probability that the event does not happen. If $\hat{p}$ is the sample proportion, the sample odds are $\hat{p}/(1-\hat{p})$. The **logistic regression model** relates the (natural) log of the odds to the explanatory variables. In the case of a single explanatory variable x, the logistic regression model is

$$\log\left(\frac{p_i}{1-p_i}\right) = \beta_0 + \beta_1 x_i$$

where the responses y_i, for i = 1,..., n, are n independent binomial random variables with parameters 1 and p_i, i.e., they are independent with distributions $B(1, p_i)$. The **parameters** of this logistic regression model are β_0 (the intercept of the logistic model) and β_1 (the slope of the logistic model). The quantity e^{β_1} is called the **odds ratio**.

278

The formulas for estimates of the parameters of the logistic regression model are, in general, very complicated and, in practice, the estimates are computed using statistical software. Such software typically gives estimates b_0 for β_0 and b_1 for β_1 along with estimates SE_{b_0} and SE_{b_1} for the standard errors of these estimates. A **level C confidence interval for the intercept β_0** is then determined by the formula

$$b_0 \pm z^* \, SE_{b_0}$$

where z^* is the upper $(1-C)/2$ quantile of the standard normal distribution. Similarly, a **level C confidence interval for the slope β_1** is determined by the formula

$$b_1 \pm z^* \, SE_{b_1}$$

A **level C confidence interval for the odds ratio e^{β_1}** is obtained by transforming the confidence interval for the slope, yielding the formula

$$\left(e^{b_1 - z^* SE_{b_1}}, \; e^{b_1 + z^* SE_{b_1}} \right)$$

To **test the hypotheses** $H_0: \beta_1 = 0$ and $H_a: \beta_1 \neq 0$ compute the **test statistic**

$$X^2 = \left(\frac{b_1}{SE_{b_1}} \right)^2$$

Since the random variable X^2 has an approximate χ^2 distribution with 1 degree of freedom, the P-value for this test is $P(\chi^2 \geq X^2)$. This is the same as testing the null hypothesis that the odds ratio is 1.

In **multiple logistic regression**, the response variable again has two possible values, but there can be more than one explanatory variable. Multiple logistic regression is analogous to multiple linear regression. Fitting multiple logistic regression models is done, in practice, with statistical software.

GUIDED SOLUTIONS

Exercise 15.7

KEY CONCEPTS - odds, odds ratio

a) You need to compute

$$\hat{p}_{high} = \frac{\text{\# in high blood pressure group who died of cardiovascular disease}}{\text{total number of men in the high blood pressure group}}$$

$$=$$

and then

$$\text{Odds} = \frac{\hat{p}_{high}}{1 - \hat{p}_{high}} =$$

b) Repeat the type calculations you did in (a), now for the men with low blood pressure.

$$\hat{p}_{low} =$$

$$\text{Odds} =$$

c) In this setting, recall that the odds ratio is

$$\text{Odds ratio} = e^{\beta_1} = \frac{\text{Odds for men with high blood pressure}}{\text{Odds for men with low blood pressure}} =$$

Interpret this odds ratio in words.

Exercise 15.9

KEY CONCEPTS - logistic regression, confidence interval for the slope, test that the slope is 0

a) A 95% confidence interval for the slope β_1 is determined by the formula $b_1 \pm z^* \, \text{SE}_{b_1}$, where z^* is the upper 0.025 quantile of the standard normal distribution.

Look up $z*$ in Table A and then, using the information given in the problem, compute

$$b_1 \pm z* \operatorname{SE}_{b_1} =$$

b) Recall that the X^2 statistic for testing the null hypothesis that the slope is zero is

$$X^2 = \left(\frac{b_1}{\operatorname{SE}_{b_1}} \right)^2 \text{ and has a } \chi^2 \text{ distribution with 1 degree of freedom. Use the}$$

values for b_1 and SE_{b_1} given in the problem to compute this quantity.

$$X^2 = \left(\frac{b_1}{\operatorname{SE}_{b_1}} \right)^2 =$$

Now use Table F to find the approximate P-value, i.e.,

$$P\text{-value} = P(\chi^2 \geq X^2) =$$

c) Now summarize your results and conclusions. What can you conclude about the probability of death from cardiovascular disease for men with high blood pressure versus that for men with low blood pressure?

Exercise 15.11

KEY CONCEPTS - confidence interval for the odds ratio

a) Recall that a level C confidence interval for the odds ratio e^{β_1} is obtained by transforming the level C confidence interval for the slope, $b_1 \pm z* \operatorname{SE}_{b_1}$ to yield the formula.

$$\left(e^{b_1 - z* \operatorname{SE}_{b_1}}, \ e^{b_1 + z* \operatorname{SE}_{b_1}} \right)$$

Referring to Exercise 15.9 (a), what is the 95% confidence interval for the slope? Hence, what is a 95% confidence interval for the odds ratio?

b) What does the interval you found in (a) tell you about the odds that a man with high blood pressure dies from cardiovascular disease relative to the odds that a man with low blood pressure dies from cardiovascular disease?

Exercise 15.22

KEY CONCEPTS - multiple logistic regression

a) You will need access to statistical software that allows you to do multiple logistic regression. SAS, SPSS, Minitab (later versions), and JMP will allow you to do multiple logistic regression. Consult your software's manual for details and explanation of the output. The output should include the value of X^2 (or z) for testing that the coefficients of both the explanatory variables are 0 and will probably also give the P-value of the test. If the P-value is not given, refer to the chi-square table in the text. The appropriate degrees of freedom in this case is 2.

b) Your statistical software should provide estimates of each of the coefficients and the standard errors of these coefficients. 95% confidence intervals can then be constructed using the formula

parameter estimate $\pm$ z*(standard error of parameter estimate)

c) What do you conclude based on your findings in (a) and (b)? Would you use both SATM and SATV to predict HIGPA, just one, or neither?

COMPLETE SOLUTIONS

Exercise 15.7

a) The proportion of men who died from cardiovascular disease in the high blood pressure group is

$$\hat{p}_{high} = \frac{\text{\# in high blood pressure group who died of cardiovascular disease}}{\text{total number of men in the high blood pressure group}}$$

$$= \frac{55}{3338} = 0.0165$$

The odds are

$$\text{Odds} = \frac{\hat{p}_{\text{high}}}{1 - \hat{p}_{\text{high}}} = \frac{55/3338}{1-(55/3338)} = \frac{55/3338}{3283/3338} = \frac{55}{3283} = 0.0168$$

b) The proportion of men who died from cardiovascular disease in the low blood pressure group is

$$\hat{p}_{\text{low}} = \frac{\text{\# in low blood pressure group who died of cardiovascular disease}}{\text{total number of men in the low blood pressure group}}$$

$$= \frac{21}{2676} = 0.0078$$

The odds are

$$\text{Odds} = \frac{\hat{p}_{\text{low}}}{1 - \hat{p}_{\text{low}}} = \frac{21/2676}{1-(21/2676)} = \frac{21/2676}{2655/2676} = \frac{21}{2655} = 0.0079$$

c) The odds ratio, with the odds for the high blood pressure group in the numerator is

$$\text{Odds ratio} = \frac{\text{Odds for high blood pressure group}}{\text{Odds for low blood pressure group}} = \frac{55/3283}{21/2655} = 2.1181$$

This tells us that the odds that, in the study group, a man with high blood pressure dies from cardiovascular disease are slightly more than twice the odds that a man with low blood pressure dies from cardiovascular disease.

Exercise 15.9

a) We are given in the problem that $b_1 = 0.7505$ and $\text{SE}_{b_1} = 0.2578$. For a 95% confidence interval, from Table A we have $z^* = 1.96$, and so

$$b_1 \pm z^* \text{SE}_{b_1} = 0.7505 \pm 1.96(0.2578) = 0.7505 \pm 0.5052$$

or equivalently, the interval (0.2453, 1.2557).

b) We compute

$$X^2 = \left(\frac{b_1}{SE_{b_1}}\right)^2 = \left(\frac{0.7505}{0.2578}\right)^2 = (2.9112)^2 = 8.4751$$

Using Table F for the χ^2 distribution with 1 degree of freedom, we estimate the P-value to be

$$P\text{-value} = P(\chi^2 \geq X^2) = P(\chi^2 \geq 8.4751)$$

which is between 0.005 and 0.0025.

c) There is strong evidence that the slope parameter in the logistic regression model is different from 0 (or equivalently, that the odds ratio is different from 1). In addition we are 95% confident that the slope parameter has a value between 0.2453 and 1.2557. Although the problem does not say so, these results, along with those of Exercise 15.7, suggest that the explanatory variable must be an indicator variable with value 1 for men with high blood pressure and value 0 for men with low blood pressure. Since a positive slope parameter implies the probability of death from cardiovascular disease increases as the explanatory variable increases, we may conclude that the data provides strong evidence that the probability of death from cardiovascular disease is higher for men with high blood pressure than for men with low blood pressure.

Exercise 15.11

a) We saw in (a) of Exercise 15.9 that a 95% confidence interval for the slope in the logistic regression model is (0.2453, 1.2557). Hence, a 95% confidence interval for the odds ratio is

$$\left(e^{b_1 - z*SE_{b_1}}, \; e^{b_1 + z*SE_{b_1}}\right) = (e^{0.2453}, e^{1.2557}) = (1.2780, 3.5103)$$

b) We are 95% confident that the odds that a man with high blood pressure dies from cardiovascular disease is between 1.2780 and 3.5103 times higher than the odds that a man with low blood pressure dies from cardiovascular disease.

Exercise 15.22

a) The results of running a multiple logistic regression model to predict HIGPA from SATM and SATV are given below. We used JMP to perform the analysis.

Response: HIGPA
Converged by Gradient
Whole-Model Test

Model	-LogLikelihood	DF	ChiSquare	Prob>ChiSq
Difference	7.11016	2	14.22031	0.0008
Full	140.55964			
Reduced	147.66980			

The information regarding the hypothesis test that the coefficients for both explanatory variables are zero is found in the row labeled "Difference" in the table above. The entry under the column labeled "ChiSquare" gives the value of the test statistic X^2. This test statistics has value $X^2 = 14.22031$ and has 2 degrees of freedom (as indicated in the column labeled "DF"). The P-value of the test is given in the column labeled "Prob>ChiSq" and is 0.0008. We would reject the null hypothesis that the coefficients for both explanatory variables are zero and conclude that at least one of the explanatory variables is helpful for predicting the odds that HIGPA is 1, or equivalently, that GPA ≥ 3.00.

b) The information for the individual parameter estimates is summarized below.

Response: HIGPA
Converged by Gradient
Parameter Estimates

Term	Estimate	Std Error	ChiSquare	Prob>ChiSq
Intercept	4.54290625	1.1617665	15.29	<.0001
SATM	-0.00369	0.0019136	3.72	0.0538
SATV	-0.003527	0.0017518	4.05	0.0441

The estimates of the coefficients for SATM and SATV are given in the table above in the column labeled "Estimate." The standard errors of theses estimates are given in the column labeled "Std Error." Recall that the general formula for a 95% confidence interval would be

parameter estimate $\pm$ z*(standard error of parameter estimate)

with z* = 1.96 for a 95% confidence interval. Using the information in the table we find the following:

95% CI for coefficient of SATM: $-0.00369 \pm 1.96(0.0019136)$
$$= -0.00369 \pm 0.00375$$
$$= (-0.00744, +0.00006)$$

95% CI for coefficient of SATV: $-0.003527 \pm 1.96(0.0017518)$
$$= -0.003527 \pm 0.003434$$
$$= (-0.006961, -0.000093)$$

c) For predicting the odds that HIGPA is 1, or equivalently, the odds that GPA $\geq$ 3.00, the results of (b) show the 95% confidence interval for the coefficient of SATV does not contain 0, hence the coefficient would be declared significantly different from 0 at the 0.05 level when SATV is added to a model containing SATM. This suggests SATV improves prediction when added to a model already containing SATM. However, the 95% confidence interval for the coefficient of SATM just barely contains 0 and we would not declare the coefficient to be significantly different from 0 at the 0.05 level when added to a model containing SATV. Since 0 is very near the upper limit of the 95% confidence interval for SATM, we might nevertheless be inclined to include SATM in the model. This is further supported by the fact that the P-value for the test that both coefficients are zero is quite small. The data suggest that using both SATM and SATV as predictors is reasonable.